Breaking the Availability Barrier II
Achieving Century Uptimes with Active/Active Systems

Breaking the Availability Barrier II

Achieving Century Uptimes with Active/Active Systems

Dr. Bruce Holenstein
Dr. Bill Highleyman
Paul J. Holenstein

AuthorHouse™
1663 Liberty Drive, Suite 200
Bloomington, IN 47403
www.authorhouse.com
Phone: 1-800-839-8640

First published by AuthorHouse 5/30/2007

ISBN: 978-1-4343-1603-5 (sc)
ISBN: 978-1-4343-1604-2 (hc)
ISBN: 978-14343-1605-9 (e)

Library of Congress Control Number: 2007903898

Printed in the United States of America
Bloomington, Indiana

This book is printed on acid-free paper.

authorHOUSE®

Dedication

This book is dedicated to our spouses,
Denise, Janice, and Karen,
for their enduring patience and support.

We also dedicate this book to Jim Gray
for his fundamental contributions to transaction
processing technology on which this book is based.
Jim, an avid sailor, has been missing at sea
since January 28, 2007.

Breaking the Availability Barrier II

Contents

Breaking the Availability Barrier II

Breaking the Availability Barrier II

Breaking the Availability Barrier II

Breaking the Availability Barrier II

Breaking the Availability Barrier II

Breaking the Availability Barrier II

Breaking the Availability Barrier II

Breaking the Availability Barrier II

Chapter 8 – Eliminating Planned Outages with Zero Downtime Migrations.................... 295

Breaking the Availability Barrier II

Breaking the Availability Barrier II

Forward

Given today's technology, [six 9s] is unachievable for all practical purposes, and an unrealistic goal.
-   Evan Marcus and Hal Stern, 2000[1]

My, how things change in just a few years! Not only are we going to talk about achieving systems with six 9s availability but also with eight 9s availability and beyond. Furthermore, we are not talking just about system availability. We are talking about application service availability. After all, following a failure of some sort, if the users of an application are being serviced in an unacceptable manner (such as experiencing excessively long response times), then the application is essentially *not available*.

If you could configure your current system to:

- provide extreme availability - MTBFs measured in centuries,
- affect only a subset of users upon a failure,
- recover from any failure in subseconds to seconds,
- lose little if any data as the result of a failure,
- eliminate planned downtime,
- achieve disaster tolerance,
- use all available capacity,
- load balance at will,
- be easily expandable,
- require no change to existing applications,
- all at little or no additional cost,

wouldn't you be interested? We think so, and that is what this book is all about. Active/active systems can and do provide these benefits today.

---

[1] Evan Marcus, Hal Stern, *Blueprints for High Availability: Designing Resilient Distributed Systems*, Wiley; 2000.

Abe Lincoln said that "it is better to remain silent and be thought a fool than to speak out and remove all doubt." At the risk of sounding foolish to some, we recognize that there are naysayers who will argue that extreme availabilities cannot be achieved. In this book we are speaking out, confident that the many examples of successful installations of active/active systems will prove us not to be fools, notwithstanding Abe.

What is "This Book"?

We referred to "this book" in the previous section. Actually, when we started to write "this book," we intended it to be the second in a series on active/active systems. However, when we finished it, it became apparent that it was much too long to be a comfortable single book to read. Therefore, we decided to break it up into two volumes.

We will refer to the (now) three volumes as Volumes 1, 2, and 3. "This book" comprises Volumes 2 and 3. The titles of the active/active trilogy are:

Volume 1: *Breaking the Availability Barrier: Survivable Systems for Enterprise Computing*, published by AuthorHouse in 2004,

Volume 2: *Breaking the Availability Barrier II: Achieving Century Uptimes with Active/Active Systems*, published by AuthorHouse in 2007 along with Volume 3.

Volume 3: *Breaking the Availability Barrier III: Active/Active Systems in Practice*, published by AuthorHouse in 2007 along with Volume 2.

In keeping with Volumes 2 and 3 being essentially one book, this Forward is the same in each volume. However, the content of each volume is markedly different.

Let us now return to the introduction of active/active systems.

Achieving Extreme Availabilities

The secret to the achievement of extreme availabilities is in the configuration. By configuring (or re-configuring) your monolithic system as an active/active architecture, the benefits described in our introduction can all be achieved.

What is an active/active system? We define it as *a network of independent processing nodes, each having access to a common replicated database. All nodes can cooperate in a common application, and users can be serviced by multiple nodes.*

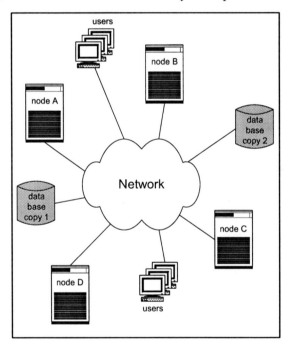

An Active/Active System

Note an important implication of this definition. Active/active architectures are not just about protecting against hardware failures. In most cases, any event that will bring down a monolithic system will only bring down one node in an active/active system. Such failure events include not only hardware faults, but also software faults, operator errors, environmental failures (air conditioning, power, etc.), and manmade or natural disasters. Active/active architectures protect users against all of these faults, allowing service to be continued by simply switching users from a failed node to one or more surviving nodes.

Another implication is what active/active is not. Active/active is not a technology; it is a business solution. Active/active is not about distributed database synchronization; it is about achieving century uptimes. More specifically,

- Active/active systems are not co-located clusters. A basic tenet of active/active systems is that they protect against area-wide problems. If the nodes cannot be geographically separated, then they are not part of an active/active system.

- Active/active systems are not independent nodes using a common database. In such an architecture, the database cannot be geographically distributed and represents a single point of failure.

- Active/active systems are not those that use hardware replication for database synchronization. Hardware replication cannot guarantee referential integrity.[2] As a consequence, applications at synchronized sites cannot use the database copies.

- By the same token, active/active systems are not those that use software replication engines that do not guarantee referential integrity.

- Active/active systems are not clusters. Users on an active/active system can be put back into service in seconds by

---

[2] See Chapter 4, Volume 2, <u>Active/Active and Related Technologies</u>.

switching them to another operating node. Clusters require that another node be brought online, a process that typically takes minutes. This time delay precludes century uptimes.

- Active/active systems are not lock-stepped or voting systems because such designs require each node to process the same requests, thus precluding scalability.

- Active/active systems are not limited to enterprise applications. There are active/active distributed database systems on the market that are loosely coupled and synchronized by replication.

- Active/active systems do not require distributed disk-resident databases. Many active/active systems maintain their databases in memory.

Of course, in some cases, there may be no need for a database in an application (for example, a cluster of Web servers). In such systems, there is no context saved between operations. Implementing clusters of systems such as these is not a difficult task as it is only necessary to route any transaction to any surviving server. However, if an active database is involved such that context is retained from transaction to transaction, then providing a redundant synchronized database is necessary. This brings with it a myriad of issues. These volumes concentrate on applications which depend upon an integrated and updatable distributed database.

In many cases, the nodes in the application network are completely symmetric. Any transaction can be routed to any node, which can read or update any set of data items in the database. Should a node fail, users at the other nodes are unaffected. Furthermore, the users at the failed node can be switched quickly to surviving nodes, with their services restored in seconds or less.

In seconds is the secret. Common today is the use of cluster technology to provide high availability. Should a node in the cluster fail, users are switched to a backup node. However, the applications on that node must be brought up and database tables and files opened before application services can be offered to the users. This process typically takes several minutes or more. In active/active configurations, all applications are already up and running on each node and are actively processing transactions. All that must be done is to switch over affected users to surviving nodes.

Let us say that an active/active system can recover services in three seconds and that the equivalent cluster can recover in five minutes (300 seconds). The cluster will be down one-hundred times longer than will the active/active system. This lops off two nines from the cluster's availability relative to the equivalent active/active system. A six 9s active/active system would be reduced to an availability of four 9s if it were in a cluster configuration. No wonder in 2007 many pundits still state that six 9s is not possible. But it is, as we will show in these volumes.

This leads to one of our availability rules:

Rule 36: *To achieve extreme reliabilities, let it fail; but fix it fast.*[3]

Are extreme availabilities important to you? Are the four 9s available with HP NonStop servers or with PC or Unix clusters acceptable? As we will discuss later, surveys have shown that the costs of downtime can range from USD \$100,000 to several million dollars an hour, depending upon the application. Perhaps even worse, downtime can lead to the dreaded "CNN Moment" and massive losses in stock value (see Chapter 9, Total Cost of Ownership (TCO), in Volume 2 for what happened to AOL in 1996 and eBay in 1999). At the extreme, downtime can lead to significant property loss or even loss of life.

---

[3] Rules 1 through 35 are formulated in Volume 1 of this series. The complete set of rules are summarized in Appendix 1 of both Volumes 2 and 3.

Only you can make this judgment. If extreme availabilities are important to your enterprise, "this book" is for you.

A Roadmap Through "This Book"

As we explained earlier, "this book" is in fact Volumes 2 and 3 of our trilogy describing how to achieve extreme availabilities with active/active systems. The first volume in this series, published in 2004 by AuthorHouse (www.authorhouse.com) and entitled *Breaking the Availability Barrier: Survivable Systems for Enterprise Computing*, referred to herein as Volume 1, lays the groundwork and the theory supporting the concepts of active/active systems. These two current volumes focus more on the practical aspects of implementing these systems.

They are broken into four parts, Parts 1 and 2 being in Volume 2 and Parts 3 and 4 being in Volume 3:

- Part 1, <u>Survivable Systems for Enterprise Computing</u>, summarizes and expands on Volume 1 and provides the background for the further topics discussed in these Volumes 2 and 3. Volume 1 is not needed to understand the content or the conclusions of Volumes 2 and 3.

- Part 2, <u>Building and Managing Active/Active Systems</u>, demonstrates how to build the redundancy required by active/active systems and how to control their cost and performance.

- Part 3, <u>Infrastructure Case Study</u>, describes an example of commercially available infrastructure products known to the authors to be suitable for production active/active systems. It also provides a valuable performance analysis tool for these products.

- Part 4, <u>Active/Active Systems at Work</u>, summarizes many of the beneficial uses of active/active systems, provides several case studies of active/active systems in use today, and describes various related technologies and issues.

The authors' intended audience for these Volumes 2 and 3 and their predecessor Volume 1 includes IT executives who feel that they must reduce the downtime of their systems, system architects and senior developers who must build these systems or modify existing systems to achieve the required availability, and operations staff who must run these systems and recover from system faults.

Part 1 – Survivable Systems for Enterprise Computing

As the French biologist Louis Pasteur said, "Chance favors the prepared mind." To prepare ourselves to understand active/active systems, Volume 1 of this series laid the groundwork for active/active systems and supported the concepts with mathematical analyses. As said earlier, Part 1 of this Volume 2 summarizes and expands upon the contents of Volume 1.

In Chapter 1, <u>Achieving Century Uptimes</u>, we talk about what is reliability and how to quantify it. We then extend these concepts to extremely reliable system configurations called *active/active* systems.

Chapter 2, <u>Reliability of Distributed Computing Systems</u>, summarizes the mathematical foundations for active/active systems. For the reader who is mathematically adverse, you will be pleased to know that the rest of this book uses minimal mathematics (except for the data replication engine performance model, which is relegated to Appendix 2). In fact, Chapter 2 can be skipped without missing the main points of the material in the following chapters.

An overview of active/active systems is discussed in Chapter 3, <u>An Active/Active Primer</u>. Here we discuss in some detail the structure and characteristics of the all-important data replication engine. We also look briefly at the various failure modes and how to recover from them as well as how to control costs of active/active architectures. These later subjects are analyzed in much greater detail in Part 2 of this volume.

Part 2 – Building and Managing Active/Active Systems

The whole rationale behind active/active systems is active redundancy, which masks failures by recovering from them so rapidly that no one notices. A similar but localized philosophy is used in HP's NonStop servers, in which critical software processes are supported by backup processes in other processors resident in the same node and ready to take over in subsecond time. Also, all databases are redundant so that disk faults are masked.

There are a variety of application network topologies that have the characteristics of active/active systems. In Chapter 4, Active/Active Topologies, examples of many of these configurations are described.

In active/active systems, the inherent redundancy includes networks, databases, and processing nodes. Chapter 5, Redundant Reliable Networks, discusses ways in which to build the reliable networks needed for data replication to provide database synchronization between distributed database copies, for heartbeats to monitor the health of the processing nodes, and for users to be switched between nodes.

Chapter 6, Distributed Databases, describes how data replication engines can be used to keep in synchronism the multiple copies of a database in the application network. It discusses issues with replication such as data collisions and loss of data following a failure. Recovery from a failed database copy and access to a viable database copy following a node or network failure are explored.

The monitoring of a processing node's health is discussed in Chapter 7, Node Failures. A node can be considered to have failed if the processing system comprising that node has failed, if its database has failed, or if it has lost connectivity to the rest of the application network due to network faults. Techniques for recovering from a node failure are discussed, including issues such as tug-of-wars and operating in split-brain mode.

A highly beneficial use of controlled failures is shown in Chapter 8, <u>Eliminating Planned Outages with Zero Downtime Migration (ZDM)</u>. Planned downtime is one of the major causes of reduced application availability. In many installations, the planned downtime required to upgrade a system or to execute other maintenance functions far exceeds unplanned downtime due to faults. In active/active systems, a node can be taken out of service purposefully with little or no impact on the users. This capability can be used to advantage to upgrade hardware, operating system software, application software, database structures, and so on. This technique also allows the capacity of the application network to be easily expanded by adding new nodes online.

Controlling the cost of an active/active system is as important as it is with any other system. However, active/active systems present an additional level of complexity. There are many ways to configure an active/active system to manage the appropriate compromise between cost, availability, and performance. As we look at different potential configurations, how do we know which contenders are the least costly? What are the factors that enter into the total cost of ownership equation? These topics are discussed in some detail in Chapter 9, <u>Total Cost of Ownership (TCO)</u>.

Part 3 – Infrastructure Case Study

In the first two parts of "this book", we describe why active/active systems can provide such high availability and how to build these systems. A set of tools is described that form a basis for the implementation of active/active systems. In Part 3, we look at a set of commercially available tools that fill the needs of active/active systems, and a performance model that can be used to gauge the effectiveness of such tools. The set of tools which are described are necessarily tools with which the authors are quite familiar but are otherwise reflective of several such tools in the marketplace.[4]

The above chapters have covered two of the three legs of the active/active triangle – availability and cost. The third leg is performance. At the heart of most active/active systems is the data replication engine, and the performance of an active/active system is directly related to this engine. In Chapter 10, <u>Performance of Active/Active Systems</u>, we create a performance model for a generic data replication engine and show how its various performance measures are affected by a variety of replication engine architectures. The mathematics behind the performance model are left for Appendix 2, <u>Replication Engine Performance Model</u>, in Volume 3.

The primary facility that is required is an appropriate data replication engine. Chapter 11, <u>Shadowbase</u>, describes the Shadowbase data replication engine that has been used in many such implementations. Shadowbase is an example of a data replication engine with a very low replication latency (the time it takes for a change that is made to a source database to be propagated to the target database). Low replication latency is important to minimize data collisions and also to minimize data loss following a failure.

In order to take a node out of service and later return it to service, it is important to have a database copy facility that can copy the contents of an active database to a node about to be put into service (or even after it has been placed into service) while the source database is being updated. Chapter 12, <u>SOLV</u>, describes such a utility. Working with Shadowbase, SOLV can efficiently make a copy of an active database even while that database is being updated. In addition, future versions of SOLV will verify that two online databases are in

---

[4] See Appendix 4, <u>Implementing a Data Replication Project</u>, in Volume 1 of this series, *Breaking the Availability Barrier: Survivable Systems for Enterprise Computing*, AuthorHouse, 2004.

synchronism and will resynchronize two active databases by repairing rows with differing content.[5]

In Chapter 13, <u>ZDM with Shadowbase</u>, we discuss the use of Shadowbase and SOLV to upgrade nodes in an active/active system without taking down the applications. With Zero Downtime Migrations, planned downtime can be completely eliminated since nodes in an application network can be upgraded without denying service to any user. Upgrades can include the hardware, operating system, applications, database, and networks, among others. In addition, ZDM can be used to add nodes dynamically into an application network to expand its capacity.

Part 4 – Active/Active Systems at Work

After learning how to build an active/active system and having seen an example of a tool set needed to do this, Part 4 looks at some actual uses of this technology in place today. It also describes some related technologies and issues.

We start in Chapter 14, <u>Benefits of Multiple Nodes in Practice</u>, by summarizing the various active/active system benefits that we have discussed in the book. These benefits include achieving extreme availability and very fast response time in the face of unplanned outages and even disasters, the elimination of scheduled downtime, the efficient use of all available processing capacity, the simplification of recovery testing, and application capacity expansion, both symmetric and asymmetric.

In Chapter 15, <u>Case Studies</u>, we look at a variety of actual uses of active/active technology. Our examples come from a wide variety of industries, including financial institutions, telecommunications, travel, web services, brokerages, plant management, and even casinos.

---

[5] Check with Gravic for availability of this feature.

Finally, in Chapter 16, <u>Related Technologies and Drivers</u>, we explore some technologies that are related to availability. They include Grid Computing, the NonStop Server Advanced Architecture, Split Mirrors, the Real-Time Enterprise, Bulletproof Storage, and Virtual Tape. We also discuss the large number of regulatory requirements that may affect your availability decisions.

Appendices

Throughout all three volumes of this trilogy, a variety of rules applicable to highly available systems have been stated. These rules are summarized in Appendix 1, <u>Rules of Availability</u>. These are annotated with volume and chapter so that their context can easily be found and studied.

Appendix 1 is contained in both Volumes 2 and 3. The remaining appendices will be found in Volume 3.

Appendix 2, <u>Replication Engine Performance Model</u>, sets forth the detailed mathematics behind the data replication engine performance model summarized in Chapter 10, <u>Performance of Active/Active Systems</u>. It also structures the resulting model into a set of tables suitable for creating an Excel spreadsheet for convenient performance calculations.

Appendix 3, <u>Regulatory Requirements</u>, summarizes the various regulatory issues that may have a bearing on the availability and operations of processing systems. These regulations are referenced in Chapter 16, <u>Related Topics and Drivers</u>.

Additionally, we asked a noted consultant in the field of highly available systems, Dr. Werner Alexi, President of CS Software, Concepts, and Solutions, GmbH, to provide his comments and critique on active/active systems. His views are presented in Appendix 4, <u>A Consultant's Critique</u>.

Authors' Notes

You may have noted that this is a long book when both volumes are considered. As Winston Churchill said, "the length of this document defends it well against the risk of its being read." To mitigate this, we would like to point out that most detail is summarized in snippets that can easily be scanned, often as rules. For instance, you might want to just hunt for the rules and read the supporting text. This will give you a good feeling for where we are trying to take you.

In many places throughout this book, reference is made to HP NonStop systems. NonStop systems were originally developed by Tandem Computers to provide very high availability. Tandem Computers was subsequently acquired by Compaq Computers, and Compaq was then acquired by HP. HP has changed the name of the Tandem systems to HP NonStop servers. The authors have considerable experience with these systems. However, concepts and recommendations presented in this book are extendable to all types of commodity systems to make them redundant, including HP Superdome, Windows Server clusters, Unix clusters, Linux servers, and IBM Parallel Sysplex systems.

Each of the chapters in this book has been written to be self-standing at the risk of some repetition. Therefore, the reader is encouraged to pick and choose the topics of interest and to read only those chapters that apply. Adequate reference is made to other chapters to suggest further reading.

Acknowledgements

All three volumes of *Breaking The Availability Barrier* have benefited from reviews by many people. We gratefully acknowledge the contributions to this volume by Mary Heck for her contributions to Appendix 3 and by Dr. Werner Alexi for his critique, published in Appendix 4. We also thank Burt Liebowitz and John Carson, whose book *Multiple Processing Systems for Real-Time Applications* provided background for this work, and Jim Gray, whose many writings fueled the fire. They and others who have influenced this volume include:

Werner Alexi, CS Software
Wendy Bartlett, HP
Victor Berutti, Gravic
Richard Buckle, Insession
Robert Cline, SunGard Securities Processing
Dan Coughlin, First Data Corp.
Michael Crispyn, Fifth Third Bank
Terry Cumaranatunge, Motorola
Dick Davis, Gravic
Giampaolo Gandini, Telecom Italia Mobile
Jeff Glatstein, SunGard Securities Processing
Jim Gray, Microsoft
Jon Healy, SunGard Securities Processing
Mary Heck, Gravic
Tom Hoffmann, Motorola
Bill Holenstein, Gravic
Denise Holenstein, Gravic
Dan Hoppmann, A. G. Edwards
ITUG Connection staff
Clark Jablon, Akin Gump
Gene Jarema, Gravic
Jim Johnson, The Standish Group
Tim Keefauver, HP
Rob Klotz, First Data Corp.
Bill Knapp, Gravic
Bob Kossler, HP
Burt Liebowitz, Consultant
Bob Loftis, HP
Mike Nemerowski, SunGard Securities Processing
Carl Niehaus, HP
Kate Noer, SunGard Securities Processing
Gianfranco Pompado, Telecom Italia Mobile

Tullio Privitera, Telecom Italia Mobile
Janice Reeder, The Sombers Group
Steve Saltwick, HP
Harry Scott, Carr Scott Software
Scott Sitler, HP
Gary Strickler, Gravic
Bart van Leeuwen, Rabobank
Joanne Welk, Motorola

About the Authors

Paul J. Holenstein is Executive Vice President of Gravic, Inc., the maker of the Shadowbase line of data replication products. Shadowbase is a low latency, high-performance, real-time data replication engine that provides business continuity as well as heterogeneous data integration and synchronization. Mr. Holenstein has more than twenty-five years of experience providing architectural designs, implementations, and turnkey application development solutions on a variety of Unix, Windows, and VMS platforms, with his HP NonStop experience dating back to the NonStop I days. He was previously President of Compucon Services Corporation, a turnkey software consultancy that was acquired by Gravic. Mr. Holenstein's areas of expertise include high-availability designs, data replication technologies, disaster recovery planning, heterogeneous application and data integration, communications, and performance analysis. He has published extensively on availability topics and is a coauthor of Volume 1 of this series. Mr. Holenstein, an HP-certified Accredited Systems Engineer (ASE), earned his undergraduate degree in computer engineering from Bucknell University and a master's degree in computer science from Villanova University. He has co-founded two successful companies and holds patents in the fields of data replication, data integration, and active/active systems. He can be reached at shadowbase@gravic.com.

Dr. Wilbur H. (Bill) Highleyman brings more than forty years experience in the design and implementation of real-time, mission-critical computing systems for companies such as Amtrak, Time, McGraw-Hill, Chemical Bank, Chicago Transit Authority, and Dow Jones Telerate. He is Chairman of The Sombers Group, a turnkey custom software house specializing in the development of real-time, online data processing systems with particular emphasis on fault-tolerant systems and large communications-oriented systems. In addition, he is the managing editor of the *Availability Digest* (www.availabilitydigest.com), a monthly periodical discussing topics in high availability with a focus on active/active systems. He was also Chairman of NetWeave Corporation, which developed the middleware product NetWeave. NetWeave is used to integrate heterogeneous computing systems at both the messaging and the database levels. Dr. Highleyman, a graduate of Rensselaer Polytechnic Institute and MIT, earned his doctorate in electrical engineering from Polytechnic Institute of Brooklyn. He has published extensively on availability, performance, middleware, and testing. He is the author of *Performance Analysis of Transaction Processing Systems*, published by Prentice-Hall, and is a coauthor of Volume 1 of this series. He holds several patents, including those in the areas of data replication and active/active systems. He can be reached at billh@sombers.com.

Dr. Bruce D. Holenstein is President and CEO of Gravic, Inc. Gravic's Shadowbase software supports many of the architectures described in this book and operates on systems such as Unix, Windows, NonStop, and other platforms running databases such as Oracle, Sybase, DB2, SQL Server, and SQL/MP. Dr. Holenstein began his career in software development in 1980 on a Tandem NonStop I. His fields of expertise include algorithms, mathematical modeling, availability architectures, data replication, pattern recognition systems, process control, and turnkey software. He is a coauthor of this series. Dr. Holenstein earned his undergraduate degree in Electrical Engineering from Bucknell University and his doctorate in Astrophysics from the University of Pennsylvania. Dr. Holenstein has cofounded and run three successful companies and holds patents in the field of data replication. He can be reached at shadowbase@gravic.com.

Part 1 – Survivable Systems for Enterprise Computing

Active/active systems are a powerful approach for achieving extraordinary application availabilities from system and networking components that have ordinary availabilities. An active/active system is a network of independent processing nodes with access to a common replicated database, all cooperating in a common application. Its primary attribute, made possible because everything is duplicated and running, is its extremely fast (subsecond to seconds) recovery time. *Let it break, but fix it fast.*

In Volume 1 of this series, *Breaking the Availability Barrier: Survivable Systems for Enterprise Computing*, we explored the theory behind active/active systems and discussed techniques for implementing them with off-the-shelf products.

In this first part, we summarize Volume 1 and set the groundwork for extending the discussion into the practical aspects of active/active systems. In Chapter 1, we explore the needs and aspects of highly reliable systems. In Chapter 2, we summarize the theory behind active/active systems (you can skip this chapter if you are adverse to mathematics – most conclusions in the book don't depend upon this material). Finally, in Chapter 3, we review the various methods for building active/active systems.

Chapter 1 - Achieving Century Uptimes

"An hypothesis is a novel suggestion that no one wants to believe. It is guilty until found effective."

- Edward Teller

The World Trade Center disaster of Sept. 11, 2001, raised the concept of business continuity to the top of corporate consciousness. The great Northeast Blackout in North America in 2003 reaffirmed how difficult it is to plan and execute a business recovery plan and how important it is to have available effective corporate data processing facilities and up-to-date corporate data at remote facilities. The Blackout went further and changed the highly available system paradigm. No longer is having systems ten miles apart sufficient. Hundreds or thousands of miles are needed to survive some disasters. If the failure of your system carries with it significant penalties in terms of cost, lost customers, unbearable publicity, or even loss of life or property, then achieving extreme reliability is of paramount interest to you.

Reliable systems do not come for free. The price of redundancy is measured both in terms of cost and in terms of performance. In fact, we can have high reliability, low cost, and high performance. Pick any two.

This second and third volumes of our series on availability[6] are focused on how you can make the best compromise between these three factors – reliability, cost, and performance - while achieving extreme uptimes measured in terms of centuries.

---

[6] W. H. Highleyman, Paul J. Holenstein, Bruce D. Holenstein, *Breaking the Availability Barrier: Survivable Systems for Enterprise Computing*, AuthorHouse; 2004; Volume 1.

What is Reliability?

Reliability is not really the reliability of a system. It must be considered the reliability of services that are provided by the system to its user community. A system may be up and running, but if it is not supporting its users in the way that they require, then for all practical purposes the system is considered *down.*

To say that a service is reliable is a qualitative statement. What might be an acceptable availability for one application may be totally unacceptable for another. We therefore need a measure of reliability[7] so that the users of a system can specify what reliability they need. This is often written into a Service Level Agreement (SLA) as a requirement for each service to be provided. When we say "users," we recognize that a user may be a person, an application, or another system.

With respect to reliability, there are two states of a service so far as a user is concerned – it is *up* or it is *down.* If it is *up,* the user is being provided with satisfactory service by the system. Satisfactory service means that the functions required by the user are operational and that the system is responsive.

A service is *down* from a user's viewpoint if one or more required functions are unavailable or are performing so poorly as not to be useful. For instance, if the user requires subsecond response time, and if the response time degrades to ten seconds, the service may be deemed to be down so far as the user is concerned.

When a service goes down, it is said to have suffered a *failure.* The measures of reliability deal with the characteristics of service

---

[7] According to our definitions, availability and reliability are complementary terms. Availability is a measure of uptime, and reliability is a measure of downtime. We use these terms loosely here, but become more formal later.

failures. As we shall see, the reliability requirements for different services may deal with different failure characteristics.

MTBF and MTR

There are two primary characteristics of failures:

(1)   How long is a service expected to be up before it fails? This is called its *mean time before failure*, or *MTBF.*[8]

(2)   How long does it take to restore the service? This is called its *mean time to repair (MTR).*[9]

In some applications, MTBF is the governing factor. A satellite, for instance, is generally not repairable. When it fails, it remains useless thereafter. In this case, only MTBF counts. MTR is not a consideration. The SLA for a satellite might be specified as an MTBF that exceeds 20 years.

In other applications, MTR is the governing factor. Short outages measured as a few seconds might be tolerable in an emergency call system such as 911 systems in the United States. A ten-second outage might be an aggravation, but a ten-minute outage could mean a life lost or a building destroyed. The SLA for a service such as this might, for example, specify that outages shall occur no more frequently than once per month (its MTBF), and that 99.9% of all outages shall be restored within ten seconds (its MTR). This means that less than 1 out

---

[8] There can be a minor confusion here in that MTBF is also used to designate mean time *between* failures, or the average time from one failure to the next. For the reliabilities that we will be discussing, these two meanings are virtually the same.

[9] We use "mean time to *repair*" in this chapter somewhat loosely, as it is the normal usage. However, we discuss the "4 Rs" in Chapter 2 – repair, recovery, restore, and return. There we redefine MTR to be the mean time to *return* the system to service. It is the sum of the repair, recovery, and restore times. MTR is more precisely the proportion of time that the system is out of service. It is the system *downtime*.

of 1,000 outages are allowed to last longer than 10 seconds – an average of one per 1,000 months, or approximately eighty years.

Other examples of services that may fail but which must be restored very quickly include nuclear power plant control systems, patient monitoring systems, and electric grid control systems.

Recovery Time Objective (RTO)

A service's MTR requirement is referred to as its *RTO*, or *recovery time objective*.

In commercial computing, the concern is often the cost of downtime. Cost may be measured in dollars, in lost customers, in bad publicity, or in a number of other ways.

For instance, a stock exchange might survive occasional brief outages for a few seconds and thereby suffer lost revenue from the loss of trades. Anything more than this could make the headlines.

Should an online store go down, sales are lost as customers find alternatives. Given enough bad experiences, the store could begin to lose its customer base.

Studies by the HP Online User Group Advocacy program[10] showed that, for over half of the respondents, downtime cost was in excess of $1,000 USD per hour. For 20% of the respondents, the cost of downtime ranged from $100,000 USD per hour to $14 million USD per day to the incalculable.

In a separate study, the July 3rd, 2006, issue of USA Today noted that 40% of 300 companies surveyed reported that their downtime

---

[10] Survey and instapoll results posted on www.hpuseradvocacy.org. on October 2, 2003, and December 18, 2003, respectively.

cost exceeded $250,000 per hour, and 14% said that their downtime cost exceeded $1,000,000 per hour (page 1B).

Availability

Thus, for commercial applications, a third reliability measure is appropriate, and this is *availability*. Availability is the percentage of time that a service is up. For instance, an availability of 99.9% means that the service will be up 99.9% of the time. It will be down 0.1% of the time, or an average of about 8.8 hours per year (there are almost 8,800 hours in one year).

This does not mean that the service will have one yearly failure that lasts for 8.8 hours. It may fail ten times a year and be down for an average of .88 hours during each failure, or it may fail once every five years and be down for an average time of 44 hours.

Table 1-1 shows the translation of an availability requirement into hours per year or minutes per month.

Nines	% Available	Downtime per Year	Downtime per Month
2	99%	87.6 hours	7.3 hours
3	99.9%	8.76 hours	44 minutes
4	99.99%	53 minutes	4.4 minutes
5	99.999%	5.3 minutes	27 seconds
6	99.9999%	32 seconds	2.7 seconds

Average 24x7 Downtime
Table 1-1

Recovery Point Objective (RPO)

There is one additional factor that must be considered, and that is the application's tolerance to data loss that might occur as the result of a failure somewhere in the application network. The amount of data that can be lost as a result of a failure is referred to as the application's *RPO, or recovery point objective.*

Systems requiring extreme levels of availability – MTBFs measured in centuries – must be redundant. This includes having duplicate database copies, which must often be distributed geographically to provide a level of disaster tolerance as shown in Figure 1-1. These databases must be kept in synchronism. When a change is made to one database, that change must be propagated to the other databases in the network.

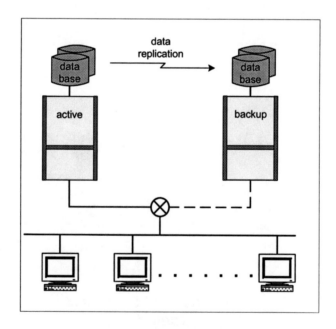

A Redundant System
Figure 1-1

It is probable that when a failure occurs, the failed system is in the midst of sending updates to other copies of the database in order to keep them synchronized. This data may be lost and may or may not be recoverable. In some applications, data lost in this manner is unacceptable.

For example, a stock exchange must never lose a transaction (stock prices would be affected for the rest of the trading day). The RPO for its trading applications must be zero.

In other applications, it may be possible to reconstruct or to ignore lost data. In these applications, seconds or even minutes of data loss may be tolerable; and the RPO may be relaxed. For instance, an ATM often keeps a local paper copy of its transactions; and a temperature monitoring system may not need to keep historical values.

System Availability

The Availability Relationship

It is clear that the availability of a service is dependent both upon its MTBF (uptime) and its MTR (downtime).[11] A service is, of course, either up or down. An up/down cycle lasts for a time of (MTBF+MTR). Therefore, its availability *A* is

$$\text{availability} = A = \frac{\text{MTBF}}{\text{MTBF} + \text{MTR}} \qquad (1\text{-}1)$$

---

[11] If you are math averse, skip the following equations; and just read the conclusions at the end of this section.

9

If MTR is very much smaller than MTBF, as is the case for the services which we are considering, then this can be written as[12]

$$A = \frac{1}{1 + \dfrac{MTR}{MTBF}} \approx 1 - \frac{MTR}{MTBF} \qquad (1\text{-}2)$$

where "$\approx$" means "approximately equal to."

The probability of a system being up (A) is one minus the probability that it is down (F):

$$A = 1 - F \qquad (1\text{-}3)$$

Thus, from Equation (1-2),

$$F = 1 - A \approx \frac{MTR}{MTBF} \qquad (1\text{-}4)$$

Finally, from Equation (1-4), we can write

$$MTBF \approx \frac{MTR}{1 - A} \qquad (1\text{-}5)$$

For instance, a service with an availability, A, of .9999 and a repair time, MTR, of four hours will have an MTBF of 40,000 hours,

---

[12] See Volume 1, Chapter 1, <u>The Nines Game</u>. As a side note, if MTBF were to be defined as mean time *between* failures, then

$$A = \frac{MTBF - MTR}{MTBF} = 1 - \frac{MTR}{MTBF}$$

or about five years. We will use this relationship to compare reliabilities in our discussions to follow.

There is one additional interpretation of availability worth noting. So far, we have characterized availability as a measure of service uptime (or service downtime). For instance, an availability of .99 means that the system is expected to be down 1% of the time, or about 88 hours per year.

Availability can also be interpreted as the probability that a user will find the system to be up when service is requested. For instance, if an ATM machine has an availability of .99, then a customer can expect that 99 times out of a hundred, the ATM will be found to be working. One out of a hundred times, he will find the ATM to be out of service.

The Reliability Relationship

For purposes of our ongoing discussions, we can now define reliability more specifically. We define reliability as the probability F that the system will be down:

$$reliability = F = 1\text{-}A$$

Thus, a system with an availability of .999 will be down 0.1% of the time and is ten times more *reliable* than a system with an availability of .99, which will be down 1% of the time.[13]

---

[13] Many contemporary definitions of reliability are functions of MTBF only. MTR is not considered. Our definition of reliability is an inverse function of MTBF for a fixed MTR. A system with an MTBF of 1,000 hours will be 10 times more reliable than a system with an MTBF of 100 hours for a given MTR.

The 9s Measure of Availability

As we have seen, availability is the probability that the system will be operational and will be providing acceptable service to its users. An availability of .999 means that it will be available 99.9% of the time.

However, for extreme availabilities, it can become quite awkward to talk about an availability of .9999999. We therefore characterize availability as the number of nines. For instance,

- 0.999 is three 9s.
- 0.9999999 is seven 9s.
- 0.998 is a little less than three 9s.
- 0.9992 is a little more than three 9s.

The number of 9s relates directly to the expected downtime for a given period of time, as was set forth in Table 1-1.

Several studies have been made of the availability of various systems in actual practice. These include all sources of failure – hardware, software, people, and environment (air conditioning, power, and so forth). A study by the Gartner Group[14] revealed the following availabilities:

HP NonStop	.9999
Mainframe	.999
OpenVMS	.998
AS400	.998
HP-UX	.996
Tru64	.996

---

[14] Gartner Group; 2002.

Solaris   .995
NT Cluster   .992-.995

**Typical System Availabilities
Table 1-2**

Thus, according to this study, mainframes are four to five times more reliable than Unix systems; and HP NonStop servers are up to ten times more reliable than mainframes.

The Price of Reliability

Reliability does not come for free. It requires a compromise with cost and performance. As we said earlier, you can optimize reliability, cost, and performance. Pick any two.

Reliability and Cost

Hardware

Reliable, highly available systems require redundancy so that if one component fails, there is another to automatically and rapidly take its place. True, today's nonredundant systems are often characterized as "highly available." These systems typically have availabilities of three 9s. But if downtimes measured in hours per year are unacceptable, then in today's technology, redundancy is a requirement.

This means that a reliable system will have extra processors (at least one), extra disks in a mirrored or RAID configuration, dual LANs, and extra power supplies and fans. In many cases, there may even be an extra system used as a spare.

Software

Software licensing may be a factor in certain cases. If the reliable architecture depends upon distributed nodes cooperating with each other, then there may be additional license fees, depending upon the licensing model of the vendor.

People and Sites

Again, if reliability involves distributing the application over multiple sites (typically done for disaster tolerance), there is the added cost for these additional sites and the staff to operate them.

Environment

A loss of power, a loss of air conditioning, or other environmental problems can take down a site. At the least, there must be backup generators to take over in the event of a power failure and with enough fuel on hand to allow time for fuel replenishment during an extended outage. Furthermore, there must be a UPS (uninterruptible power supply) system with sufficient capacity to power the systems until the backup generators can come online.

In the extreme, the Northeast Blackout demonstrated that distributed nodes should not only be serviced by independent power suppliers but also should be located in areas serviced by different power grids.

Even mundane things like water must be considered. Not only should water be available for staff and perhaps air conditioning, but the lack of water will make toilets inoperable.

Networking

If the system depends upon a network to connect users or to coordinate nodes in a distributed application, then the network should also be redundant. As illustrated by the 9/11 terror attacks, these networks should not only follow different paths but ideally should be provided by different carriers to protect against a carrier outage.

Reliability and Performance

Reliability can also take its toll on the response time and capacity of a system.

Software

Like hardware, software must be redundant because it, too, can fail. If a software module fails, then there must be another module to take its place – either a spare module or a currently active module that can assume the load of the failed module.

In some cases (like HP NonStop process pairs), each critical process[15] has a backup process. The backup's context is kept in synchronization with the primary process via a mechanism known as checkpointing. Checkpointing imposes a nontrivial load affecting both application responsiveness and system capacity.[16] Its virtue is that recovery from a software fault is almost instantaneous and can be transparent to the application.

Another technique is the use of *persistent processes*. Critical processes are monitored by a resilient process (perhaps a check-

---

[15] A process is a program running in a processor.
[16] W. H. Highleyman, Chapter 9, <u>Fault Tolerance</u>, *Performance Analysis of Transaction Processing Systems,* Prentice-Hall; 1988.

pointed process pair). Should such a critical process fail, the monitor will reinstantiate it in a surviving processor.

Yet another technique is to provide a group of context-free processes that can handle a particular class of transactions, as do BEA's Tuxedo, IBM's CICS, and HP's NonStop Pathway. If one process fails, it is either restarted, or it is removed from the group with the remaining processes sharing the load. This technique imposes less overhead on the system, but failover can take seconds rather than subseconds – a tradeoff between capacity and response time.

Transaction Management

It is imperative to keep the system's database consistent if it is to be reliable. This task falls to transaction managers such as IBM's IMS and CICS, Tuxedo's distributed transaction manager, and HP's NonStop TMF.

With a transaction manager, a set of database operations representing a logical or business transaction is marked as an atomic set. The transaction manager guarantees either that all operations for the transaction are completed or that none are and ensures that any view of the database is consistent during the processing of the transaction.

A transaction manager may impose a significant performance penalty in terms of capacity, concurrency, and response time.

Redundant Disks

To achieve high reliability, disks must be redundant. This means additional disk write activity as the redundant disks are updated. Mirrored disks must be written sequentially to avoid data loss due to a failure during the write process. This can significantly increase both capacity utilization and response times.

RAID arrays can impose less overhead because writes to the several disks are done in parallel. However, a failure during the write process can cause data contamination.

In general, with redundant disk systems, there is the opportunity to improve read performance by spreading the read activity over the several disks. Many applications are characterized by disk write-once, read-many operations and can significantly benefit from this approach.

Node Synchronization

The achievement of extreme reliability often means that the application is distributed among several independent but cooperating nodes. This requires that nodes be cognizant of activity at all other nodes. More specifically, each node must have access to a common database that holds the current state of the application. The database must be redundant, with redundant copies distributed to survive node outages. The various copies of the database across the network must be kept in synchronism so that anyone accessing the data from any copy will get the same results.

Keeping the databases of these nodes in synchronism will impose additional capacity utilization on all nodes. Depending upon how tightly the synchronization must be maintained, response times can be adversely affected as well.

Synchronization considerations are explored in some detail in Chapter 3, An Active/Active Primer.

The Why of Century Uptimes

Why the quest for uptimes measured in centuries? Even if we thought that we could achieve this, could we ever confirm it by field experience? Is it really meaningful? After all, we will all be gone; and the system will be retired long before a century is up.

The answers to these questions are "yes, we can measure century uptimes" and "yes, century uptimes are meaningful." Let us look first at measuring these long uptimes.

Some applications now running require a hundred or more systems (online stores and Internet services such as email and search engines are but one example). If such an application requires one hundred systems, and each has a one-century uptime, then we can expect on average one system failure per year.

Furthermore, if in the future a vendor successfully markets extremely reliable systems with, say, ten-century uptimes and has an installed base of 2,000 systems, one would expect this community of systems to experience two outages per year. So it is perfectly reasonable to assume that we can measure this extreme reliability in many cases.

But is this extreme reliability useful? Let us take the case of an extremely reliable system with four 9s availability. This corresponds roughly to the best of today's commercial offerings. It means that the probability of failure is 0.01%.

Put another way, it means that someone occasionally using the services of this system will find it unavailable on the average of one time out of every 10,000 times that service is sought.[17] If this results in a simple inconvenience, then the level of availability is probably more than sufficient. If it means the possibility of a nuclear meltdown, the loss of life, or the death of a $100 million satellite, then more 9s are obviously needed. If it means the loss of $100,000 per hour or bad publicity, then adding more 9s might be strictly a matter of cost. (The cost of reliability is explored further in Chapter 3, An Active/Active Primer, and in Chapter 9, Total Cost of Ownership (TCO).)

Thus, there are many cases in which extreme reliabilities measured in centuries merit strong consideration. In these cases, we must be prepared to cope with any failure, even if we don't know what it will be.

Besides, an uptime of a century means that someday in the next century the system will likely fail. That someday may be tomorrow.

The How of Century Uptimes

Doubling Your 9s

It is shown in Volume 1 that adding a second redundant system to a single system doubles its 9s.[18] If, as in our earlier Figure 1-1, we provide a Unix system that has three 9s of availability with an equivalent standby system that can assume control transparently to the users in the event of a primary system failure, we now have a system with six 9s of availability. This certainly qualifies as an extremely available system, having an expected downtime of only about 30 seconds per year. Assuming a repair time (MTR) of four hours (a generally accepted field service interval), this represents an MTBF of 5 centuries.

---

[17] Though this statement is true, the reason that we say "occasionally" is that if a service is used frequently (say by another process every second), and if the system is down for four hours, the result will hardly be an inconvenience.

[18] See Volume 1, Chapter 1, The Nines Game.

To understand this simple rule is straightforward. Using the above example, the probability of random failure of a system with three 9s is .001. If we provide a similar redundant system, then a system failure occurs only if both systems fail. This will occur with a probability of .001 x .001 = .000001. The availability of the redundant system is therefore 1-.000001 = .999999, or six 9s.

The active/backup[19] architecture shown in Figure 1-1 is no longer state-of-the-art technology for achieving extreme reliabilities. This approach can provide very high availabilities. However, it has several problems associated with it:

- Only half of the purchased capacity is used at any one time.
- All users are affected by a primary failure and must be switched to the standby node.
- Recovery can take minutes to hours as partial transactions are backed out, as lost transactions are recovered, as applications are brought up, and as users are switched over.
- Transactions in the process of being replicated at the time of failure are generally lost.

Wouldn't it be nice if we could have an architecture that

- provided extreme availability - MTBFs measured in centuries,
- affected only a subset of users upon a failure,
- recovered from any failure in subseconds to seconds,
- lost little if any data as the result of a failure,
- eliminated planned downtime,
- achieved disaster tolerance,
- used all available capacity,
- load-balanced at will,

---

[19] The active/backup architecture goes by many names including active/standby, active/passive, primary/secondary, primary/backup, and so on.

- was easily expandable,
- required no change to existing applications,
- all at little or no additional cost?

The technology is here today to substantially meet all of these goals, and it is the active/active architecture which we will describe in detail in this volume.

The n+1 Solution

Providing a redundant backup system is the classic approach to extreme availability in older technology, but it is an expensive approach. One must purchase twice the equipment required to run the application. A more cost-effective approach is to add partial redundancy.

This is the approach taken by HP's NonStop servers. Each of these servers can contain up to sixteen processors. One configures the system with the number of processors needed (n) plus one additional one (thus the term "$n + 1$").

Normally, all $n+1$ processors are used to carry the application load. However, if a processor fails, the remaining n processors are sufficient to carry the load.

These systems are designed to survive any single hardware or software failure and many cases of multiple failures. If we assume that each processor (and its associated peripheral gear) is equivalent to our previously mentioned Unix box so far as availability is concerned, then each processor has an availability of three 9s. The probability that a specific pair of processors will fail is equivalent to an availability of six 9s, as described earlier.

However, a dual failure may potentially cause a system outage. We call this a *failure mode*. In a NonStop system, for instance, the

21

failure of two specific processors may bring down the system if a critical process pair should happen to be running in that processor pair. In a fully configured NonStop system, there are sixteen processors. Thus, there are 120 ways in which two processors can fail.[20] Let us assume that 100 of these combinations will take down the system (that is, there are 100 potential failure modes for this system).

Thus, the failure rate is 100 times worse than one would assume based on an availability of six 9s. This, in effect, reduces our availability by two 9s, leaving a typical NonStop system with four 9s of availability.

Four 9s availability is an order of magnitude better than any other system being marketed as of this writing, and this availability is borne out in industry studies we described earlier. (Of course, things are more complex than this simple analysis because of faults caused by operator errors, environmental faults, and so forth;[21] but our analysis serves to illustrate the point of this discussion.)

Dual Redundancy for Double 9s

Four 9s is a fairly high availability, but it does not come close to achieving century uptimes. If the system repair time is four hours, this equates to an average uptime of five years – impressive but not our goal.

What are our goals? As a straw man, let us say that our objectives are to provide application uptime of six 9s or better at little or no

---

[20] Specifically, $n(n-1)/2$ ways. One of n processors may fail, followed by one of the remaining $n-1$ processors. But this has counted each processor pair twice – for instance, processor 3 followed by processor 4 and processor 4 followed by processor 3.

[21] See Volume 1, Chapter 5, <u>The Facts of Life</u>, and in this Volume, Chapter 2, <u>Reliability of Distributed Computing Systems</u>.

additional cost. In addition, in order to control costs, we want to be able to use all available capacity of the components that are up. Impressive ambitions, and ones which are achievable today, as we shall see.

As we have seen from the examples above, in order to achieve this level of reliability short of buying an entire standby system, we really need two levels of redundancy. The ability to survive any single failure can in today's technology give us four 9s. To do better than that means that we have to be able to survive at least two failures.

By the way, if we assume a four-hour repair time, six 9s equates to an MTBF of five centuries. If we can achieve this goal, we have achieved century uptimes.

Stateless Distributed Systems

There is a special case for which the goal of six 9s or better is trivial, and that is when the application is stateless, or context-free. In a stateless system, the application responds to user requests independently of any prior activity. The results of previous requests have no bearing on the result of the current request. The system's response to a specific request will always be the same.[22]

An excellent example of a stateless system is a web service that delivers fixed pages of information such as catalog pages for an online store. Each request results in a page being sent that is independent of any prior request (though the requested page may have come from a link embedded in a prior page), or the request itself carries any needed context.

---

[22] This, of course, ignores the problem of rolling updates to the servers.

Rather than having one large web page server, whose failure would bring down the store's catalog, several smaller, independent servers may be used, as shown in Figure 1-2, each capable of delivering any requested page. These servers may be collocated or may be geographically dispersed to provide disaster tolerance. Each user request is routed to an available server, which will respond to the request. A user's subsequent request will most likely be routed to another server.

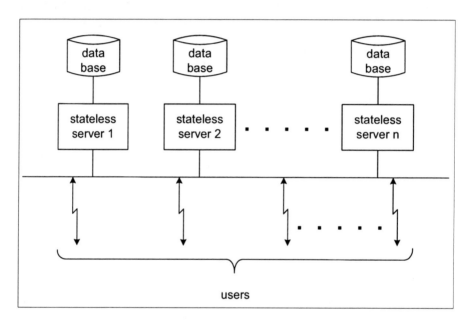

A Stateless Distributed System
Figure 1-2

Should a server fail, it is simply removed from the pool of available servers. This architecture can survive multiple server failures – up to the point where the surviving servers can no longer handle the load.

The availability analysis for stateless distributed systems is similar to that for the NonStop server given earlier, except that it is trivial to provide more than one spare. Two, three, or more spares may be provided as desired with no increase in complexity.

For instance, let us assume that we need ten servers to handle our peak load; and we provide dual redundancy via two spares for a total of twelve servers. In order for the system to fail, we would have to lose three servers (and it would still be operational except for reduced capacity). If each server has an availability of three 9s, then by our previous argument the probability of losing three servers is $(.001)^3$, or nine 9s. However, as pointed out in the NonStop case, there are 220 different ways that three servers out of twelve can fail.[23] This causes a loss of two and a half 9s, resulting in an availability of full capacity in excess of six 9s.

Thus, we have achieved an availability commensurate with providing a standby system but at an expense of 20% excess capacity rather than 100% excess capacity.

Stateful Distributed Systems

Using stateless servers is all well and good for the web-page catalog operation, but an online store is not of much use unless customers can make purchases as well. (Gone are the days when customers browsed the catalog online and then called a toll-free number to place an order.) But ordering is hardly context-free. As the customer shops, the system must maintain her shopping cart, obtain billing and shipping information, validate her credit source, confirm the order to the customer, obtain her acknowledgement, and then ensure that the order is shipped.

---

[23] Specifically, $n(n-1)(n-2)/6$, where $n = 12$.

Ordering and order fulfillment are an example of what we call a *stateful* system.

Building reliable, stateful distributed systems is our real focus in this series. *A stateful system is one in which the processing of each request is dependent upon the results of previous requests.* That is, the system must carry the context of a thread of requests and must remember all pertinent information as the transaction progresses. As a consequence, the stateful system must appear as a single system to the user with a single database of record.[24] The question, then, is how to build such a system with the additional sparing needed to achieve our reliability goals?

The answer is to take a lesson from stateless distributed systems and build a distributed system of independent cooperating nodes. The operative word here is *cooperating*. The nodes must each cooperate equally in the application and must maintain a common system state across all nodes. When a transaction occurs at one node, the results of that transaction must be communicated to all other nodes to update their state.

Active/Active Systems

What is an Active/Active System?

Typically, the system state in commercial applications is represented by the contents of the database.[25] Therefore, this cooperation means that any node can change any row in the database; and that redundant copies of the application database across the network must receive these changes in order to keep all databases synchronized.

---

[24] See Chapter 6, <u>Distributed Databases</u>, for a further description of the database of record.

[25] Even though we talk primarily about disk-resident databases, databases may be memory-resident or stored on other media.

Such systems are called *active/active* systems since all nodes in the network are actively participating in the processing of user requests. A general active/active system is shown in Figure 1-3. The distributed copies of the database are kept in synchronism via data replication across their interconnecting network.

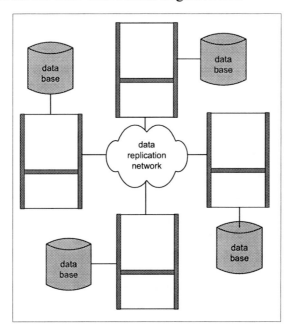

A Stateless Distributed System
Figure 1-2

Active/active systems do not protect only against hardware failures. They protect against anything that can take a node down including hardware faults, software faults, operator errors, environmental faults (power, air conditioning, etc.), and manmade or natural disasters.

Typically, the sum total of the capacity of the many nodes in the application network equals the capacity requirements of the

27

application plus some additional capacity required to support the data replication overhead necessary to keep the databases in synchronism and to provide the required capacity in the event of a node failure.

Note that in an active/active system, all of the purchased capacity is usable. There is no need for an idle standby system. In addition, disaster tolerance comes for free if the nodes are geographically distributed. This leads to the following rules:

Rule 37: *All of the purchased capacity in an active/active system is usable. There is no need for an idle standby system.*[26]

Rule 38: *Providing that the nodes in an active/active system are geographically distributed, disaster tolerance comes for free.*

There have been several precursors to active/active systems in the search for extreme availabilities, such as:

- *active/backup systems*, in which a second system is ready and able to take over processing should the primary system fail. These systems suffer from a time delay due to management involvement in the switchover decision followed by an extended time to switchover. Also, many of these systems use replication techniques which do not guarantee referential integrity, leaving the database copies unsuitable for transaction processing.

- *clusters*, which require that another node be brought online, a process that typically takes minutes.

- *multiple nodes sharing a common database*. Nodes generally are not geographically separated for disaster tolerance.

---

[26] Note that Rules 1 through 35 were formulated in Volume 1 of this series. Rule 36 is given in the Forward to this volume.

Furthermore, the common database cannot be geographically distributed.

- *lock-stepped systems*, in which two systems process the same data and compare their results on an operation-by-operation basis.

- *redundant systems*, in which data is processed independently and the results compared (this includes voting systems of three or more systems).

- *federated systems*, in which a common but non-redundant view of data distributed about the network (the *federated database)* is provided to any user no matter to which node he is connected.

None of these are active/active systems by our definition since the various systems are not independently processing transactions against a common replicated database in a loosely-coupled redundant environment. In all cases, the capacity of the entire complex is the capacity of a single system or database. Furthermore, just because bidirectional replication may be used does not make the system active/active if the replication engine does not guarantee the referential integrity of the target database, since that database cannot be used for active transaction processing.

Sparing is the number of additional redundant components that are available for taking over the functions of a failed component. As we have seen, in order to achieve extreme availabilities, we need to have at least two levels of sparing in an active/active system. If the system can withstand a single failure (as with NonStop systems), we can achieve availabilities of four 9s. If a system can survive dual failures, we can achieve availabilities in excess of six 9s.

One can achieve dual sparing in an active/active system by providing two spare nodes, neither of which is fault-tolerant. Another approach is to use fault-tolerant nodes, such as HP's NonStop servers, to implement the first level of sparing and then to provide a single spare node in the network to implement the second level of sparing. Let us look at these two approaches.

Dually Redundant Uniprocessors

Consider a Unix system that has an inherent availability of three 9s, which we would like to reconfigure for century uptimes. One way to do this, as shown in Figure 1-4, is to provide an equivalent standby system. We have doubled our cost, but we have achieved our availability goal. However, we have had to buy 200% of our required capacity; and only half of it can be used at any one time.

We can rearchitect this system by splitting it up into, say, four smaller cooperating nodes in an active/active configuration, with each node capable of carrying 25% of the load. We then add one spare system for fault tolerance. The system will continue to provide 100% capacity in the presence of any single node failure but will provide reduced capacity should two nodes fail. The result is that we have purchased only 125% of the required capacity and have achieved full-capacity availability of five 9s, or a 50-year MTBF based on a four hour MTR. Impressive, but not our goal.

In order to achieve our objective of century uptimes, we must add a second spare to become dually redundant. Now the system can survive the failure of any two nodes and has achieved a 100% availability of almost eight 9s. Thus, we have had to purchase only 150% of the required capacity and have achieved hundreds of centuries of uptime. Furthermore, all of this excess capacity is available during normal operations.

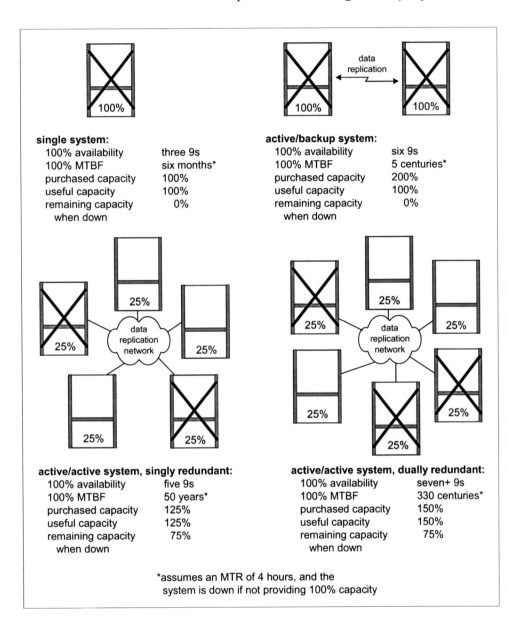

Uniprocessor Stateful Distributed Systems
Figure 1-4

Dually Redundant Multiprocessors

Alternatively, we can achieve the first level of redundancy by using fault-tolerant nodes such as NonStop servers. The second level of redundancy is provided by a single spare node. Let us assume that the application requires sixteen NonStop processors-worth of processing capacity. Figure 1-5 shows the characteristics of this system as a stand-alone system; in an active/backup configuration; and as an active/active system with five fault-tolerant nodes, each with four processors – the required configuration plus 25% excess capacity.

So far as the five-node active/active system is concerned, if each fault-tolerant node has an availability of four 9s, then the availability of any pair of nodes is eight 9s. However, there are ten ways in this system that we can lose two nodes (5x4/2). Therefore, our system availability is seven 9s, well exceeding our goal of six 9s. Seven 9s is an average of less than four seconds of downtime per year. If the recovery from a dual failure requires four hours, we have achieved uptimes in the order of 50 centuries!

In the above example, we have achieved this reliability with 20 processors instead of the 32 processors which would have been required in an active/backup configuration. In addition, in normal operation, we have available to the application 125% of the capacity of an active/backup system and still have 75% capacity following a two-node failure.

This example also shows why in some cases the extra cost of inherently fault-tolerant systems may, in fact, be cost effective. Comparing Figures 1-4 and 1-5, we see that typically it requires one less node to achieve a particular level of availability if fault-tolerant nodes are used. This leads to the following rule:

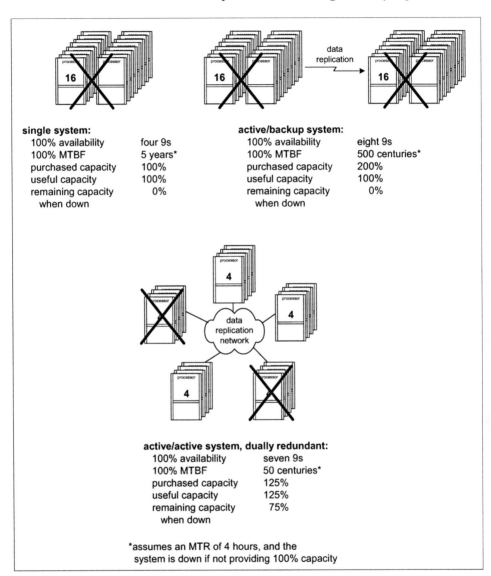

single system:
100% availability	four 9s
100% MTBF	5 years*
purchased capacity	100%
useful capacity	100%
remaining capacity when down	0%

active/backup system:
100% availability	eight 9s
100% MTBF	500 centuries*
purchased capacity	200%
useful capacity	100%
remaining capacity when down	0%

active/active system, dually redundant:
100% availability	seven 9s
100% MTBF	50 centuries*
purchased capacity	125%
useful capacity	125%
remaining capacity when down	75%

*assumes an MTR of 4 hours, and the system is down if not providing 100% capacity

A Multiprocessor Stateful Distributed System
Figure 1-5

Rule 39: (NonStop Maxim) - *Century uptimes in an active/active system will generally require fewer nodes if fault-tolerant nodes are used rather than high availability nodes.*

The Acceptance of Active/Active Technology

The distributed active/active systems that we have described can substantially achieve the lofty goals that we set:

- They do achieve century uptimes.
- They do make available for work the purchased capacity of all of the nodes in the system.
- When a node failure occurs, only users on a failed node are affected, and they can be put back into service within seconds by switching them to a surviving node.
- As we shall see later, by using synchronous data replication, we can ensure that no data is lost due to a node or network failure.
- As an added bonus, we do not need to purchase an additional 100% of required capacity. We only need to purchase a fraction of that.

Until recently, there have been significant hurdles to implementing active/active systems. The primary hurdle has been the requirement to synchronize database copies across the network, especially when multiple systems can be updating the same data at the same time.

Other hurdles include the resynchronization of a downed database after its node has been returned to service, the introduction of expansion nodes into the system, the handling of network failures, the switching of users from a downed node to a surviving node, and the need for planned downtime for software and hardware upgrades.

As described in later chapters in these Volumes 2 and 3, solutions for all of these problems now exist; and active/active systems are being deployed into service for large-scale, mission-critical applications. These systems are being deployed as leading edge applications. As they prove their worth, active/active systems will become more predominant and will replace lesser technology to become the de facto architecture of choice.

Several case studies of existing active/active systems are presented in later chapters.

High availability gives a competitive advantage to an enterprise:

Rule 40: (Darwin's extension to Murphy's Law) - *Eventually, a disaster will befall every enterprise; and only those that are prepared will survive.*

What's Next

This volume is about building cost-effective, extremely reliable systems, those exceeding six 9s of availability with uptimes measured in centuries. To achieve these goals, such systems may incur only a fraction of the cost of active/backup systems in common use. This volume focuses on the why and how of active/active systems and extends the fundamental concepts set forth in Volume 1 of this series.

The next two chapters summarize and expand the concepts presented in Volume 1. Chapter 2 analyzes why active/active systems can be so reliable. As Abraham Lincoln said, "For those who like this sort of thing, this is the sort of thing they like." If you are one who does not like math, you can skip Chapter 2. Chapter 3 summarizes the technology available today to build these systems.

Later chapters then delve into advanced topics beyond those presented in Volume 1, and present a variety of case studies describing successful implementations of active/active systems.

Chapter 2 - Reliability of Distributed Computing Systems

"He who loves practice without theory is like the sailor who boards ship without a rudder and compass."

- Leonardo da Vinci

In Chapter 1, we demonstrated how redundancy significantly improves availability and allows the system to continue to operate in the presence of one or more component failures. We explored singly redundant systems that can survive any single failure and dually redundant systems that can survive any pair of failures.

By example, we showed that singly redundant systems can typically achieve reliabilities measured in years; and dually redundant systems can achieve reliabilities measured in centuries. The examples from Chapter 1 are summarized in Table 2-1, which shows typical availabilities for standalone systems, active/backup pairs, and active/active systems for two cases:

- The system nodes are nonredundant (like Unix or Windows systems), with availabilities in the order of three 9s.

- The system nodes are redundant (like HP's NonStop systems) with availabilities of four 9s.

In both cases, active/backup pairs provide extraordinary availabilities measured in centuries. However, this architecture requires the purchase of twice the needed capacity; and half of it generally goes idle.

On the other hand, active/active systems can achieve the same availabilities, but in many cases they can be implemented with the purchase of much less capacity, thus potentially providing significant cost savings.

Breaking the Availability Barrier – Volume 2

Configuration	nonredundant nodes (e.g., Unix)	redundant nodes (e.g., NonStop)
single system (1 node total), single node failure:		
purchased capacity	(a)	(b)
	1.00	1.00
usable capacity	1.00	1.00
9s	3	4
MTBF	0.5 years	5 years
Usable capacity after failure	0.00	0.00
active/backup (2 nodes total), dual node failure:		
purchased capacity	(c)	(d)
	2.00	2.00
usable capacity	1.00	1.00
9s	6	8
MTBF	5 centuries	500 centuries
Usable capacity after failure	0.00	0.00

Reliability Examples
Table 2-1a

Configuration	nonredundant nodes (e.g., Unix)	redundant nodes (e.g., NonStop)
active/active – one spare node (5 nodes total), dual node failure:		
purchased capacity	(e) 1.25	(f) 1.25
usable capacity	1.25	1.25
9s	5	7
MTBF	50 years	50 centuries
Usable capacity after failure	0.75	0.75
active/active – two spare nodes (6 nodes total), dual node failure:		
purchased capacity	(g) 1.50	(h) 1.50
usable capacity	1.50	1.50
9s	4+	6+
MTBF	30 years	30 centuries
Usable capacity after failure	1.00	1.00

Reliability Examples
Table 2-1b

Configuration	nonredundant nodes (e.g., Unix)	redundant nodes (e.g., NonStop)
active/active – two spare nodes (6 nodes total), triple node failure:		
purchased capacity	(i)	(j)
usable capacity	1.50	1.50
9s	1.50	1.50
	8	10+
MTBF	330 centuries	50,000 centuries
Usable capacity after failure	0.75	0.75

Notes: (1) Active/active systems based on four nodes for 100% capacity.
(2) MTR of 4 hours

**Reliability Examples
Table 2-1c**

Active/Active Systems Reviewed

In the search for highly reliable solutions that were more economical than active/backup configurations, we introduced active/active systems. An active/active system is *a network of independent processing nodes, each having access to a common replicated database. All nodes can cooperate in a common application, and users can be serviced by multiple nodes.*

As we can see in Table 2-1, a singly redundant active/active system can be built from nonredundant systems by providing one extra node in the application network [Case (e)]. Alternatively, the same result can be achieved by using fault-tolerant systems which are themselves singly redundant [Case (b)]. In these examples, singly redundant systems can achieve availabilities of four or five 9s which, with a repair time (MTR)[27] of four hours, equates to MTBFs measured in years.

However, if we use dual redundancy, we can achieve availabilities of seven to eight 9s, which equate to century MTBFs. A dually redundant system can be constructed from nonredundant systems by providing two spare nodes [Cases (g) and (i)] or from fault-tolerant systems by providing one spare node [Case (f)] (see Rule 39, the NonStop Maxim, in Chapter 1). Furthermore, if two spare fault-tolerant nodes are provided (triple redundancy), 100% capacity is ensured even in the event of a dual-node failure [Cases (g) and (h)]. (Table 2-1 is extended to compare the case of two spare redundant nodes.)

---

[27] We continue here to use the usual definition of MTR as mean time to repair. We later redefine MTR as mean time to return to service, as repair is just one component of MTR. In any event, MTR is the proportion of time that the system is down (its downtime).

41

Not only can active/active systems achieve the availabilities provided by primary/standby system pairs, but they can do so with far less equipment and cost. Furthermore, all spare systems can be active during normal operations. There are no idle standby nodes in an active/active system.

Adding to our rules set forth in Volume 1 of this series, *Breaking the Availability Barrier: Survivable Systems for Enterprise Computing*, and in Chapter 1 of this volume,[28] we have:

Rule 41: *Active/active systems can provide the availability of a primary/standby pair with less equipment and less cost.*

Also, we repeat Rule 37 from Chapter 1 of this volume:

Rule 37: *All of the purchased capacity in an active/active system is usable. There is no need for an idle standby system.*

An additional advantage of active/active systems can be observed by looking at the failure modes in Table 2-1 The assumption in this Table is made that if node failures reduce the system capacity to less than 100%, the system has failed. For active/backup configurations, the system is either providing 100% capacity or 0% capacity, so that this is certainly true. However, for active/active configurations, only the capacity of one node has been lost when the system capacity falls below 100%. Though the system has failed by definition, the system still has processing capacity remaining. In the examples of Table 2-1, the system continues to provide 75% capacity following a "system failure." In some cases, this may be sufficient to continue processing, especially if non-critical load can be shed. Consequently, in these cases, the definition of system failure might be relaxed.

---

[28] See Appendix 1, <u>Rules of Availability</u>.

The Availability Relationship[29]

As discussed in Chapter 1, the reliability of a system is defined to a great extent by its MTBF and MTR:

> MTBF is the mean (average) time before a system will fail.
>
> MTR is the mean (average) time to repair a system and to return it to service.

From these can be determined the system availability, A:

$$A = \frac{MTBF}{MTBF + MTR} \qquad (1\text{-}1)$$

which can be rewritten as

$$A = \frac{1}{1 + \dfrac{MTR}{MTBF}} \approx 1 - \frac{MTR}{MTBF} \qquad (1\text{-}2)$$

where the approximation is valid if MTR is very much less than MTBF.

Since the system can only be up or down, the probability that it is down, F, is

$$F = 1 - A \approx \frac{MTR}{MTBF} \qquad (1\text{-}4)$$

and

---

[29] The rest of this chapter deals with the mathematical theory behind active/active systems. If you are math averse, feel free to move on to Chapter 3 as this material is not needed for an understanding of the following material in the book.

$$\text{MTBF} \approx \frac{\text{MTR}}{1 - \text{A}} \qquad (1\text{-}5)$$

Equation (1-5) is the relation that we used extensively in Chapter 1 to relate MTBF to availability.

Availability of Computing Networks

Let us now consider the availability of a computing system comprising a number of similar components. Examples that we have already used are an active/active system comprising several like nodes, and a NonStop fault-tolerant system comprising several like processors.

We will use capital letters for system reliability factors and lower case letters for component reliability factors. Thus,

A	is the availability of the system.
F	is the probability that the system will fail.
MTBF	is the mean time before failure for the system.
MTR	is the mean time to return the system to service following a system failure.
a	is the availability of a component.
mtbf	is the mean time before failure for a component.
mtr	is the mean time to return a component to service following a component failure.

Similar to the definition of system availability, A, as given by Equation (1-2), component availability a is related to component mtbf and mtr by

$$a \approx 1 - \frac{\text{mtr}}{\text{mtbf}} \qquad (2\text{-}1)$$

Let us start by considering the simple case of a system made up of two components (such as an active/backup pair, an active/active system with two nodes, or a NonStop system with two processors). The nodes, processors, or members of a pair are the components.

In this case, it is clear that this is a singly redundant system. There is only one spare. The system can survive the failure of one component, but it fails if both components fail.

The probability that a component will fail is $(1-a)$. The probability that the second component will fail while the first component is down is $(1-a)^2$. This is the probability, F, that the system will fail:

$$F \approx (1-a)^2 \qquad (2\text{-}2)$$

and its availability is

$$A = 1 - F \approx 1 - (1-a)^2 \qquad (2\text{-}3)$$

Let us now extend our case to a singly redundant system with more than two components. Figure 2-1 shows an example of a system with four components, C1 through C4. Let

n = number of components in the system.

Being singly redundant, the system can survive if it has $(n-1)$ components operational. However, it will fail if two components fail, leaving only $(n-2)$ operational components.

The failure of any particular pair of components will occur with a probability of $(1-a)^2$. However, there are now many ways in which two components can fail. Figure 2-1 shows that there are six ways in which two components out of four can fail. Any one of the n

components can fail, followed by any of the $(n-1)$ remaining components. Thus, there are $n(n-1)$ ways in which two components can fail. However, this has counted each component pair twice (for instance, component 2 followed by component 4 as well as component 4 followed by component 2). Thus, the number of ways that two components can fail, ignoring failure order, is

$$\frac{n(n-1)}{2}$$

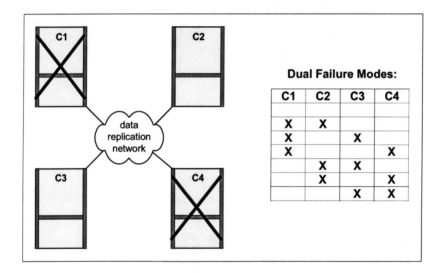

Four-Node Singly Redundant System
Figure 2-1

We call each of these pairs a *failure mode.*

Let us define *f* as the number of failure modes that will take down the system. Then, for single redundancy,

$$f = \frac{n(n-1)}{2} \qquad\qquad (2\text{-}4)$$

Since each failure mode occurs with probability $(1\text{-}a)^2$, then the probability of system failure is

$$F \approx f(1-a)^2 \qquad \text{(singly redundant)} \qquad (2\text{-}5)$$

and the system availability is

$$A \approx 1 - f(1-a)^2 \qquad \text{(singly redundant)} \qquad (2\text{-}6)$$

Let us now extend this to dual redundancy. Here, three components must fail to take down the system. This will happen with a probability of $(1\text{-}a)^3$.

There are $n\,(n\text{-}1)(n\text{-}2)/6$ ways for three components out of n to fail (one of n, then one of $(n\text{-}1)$, then one of $(n\text{-}2)$, having counted each triple six times). Thus, following the singly redundant case, we have

$$f = \frac{n(n-1)n-2)}{6} \qquad \text{(dually redundant)} \qquad (2\text{-}7)$$

$$F \approx f(1-a)^3 \qquad \text{(dually redundant)} \qquad (2\text{-}8)$$

$$A \approx 1 - f(1-a)^3 \qquad \text{(dually redundant)} \qquad (2\text{-}9)$$

In the general case, there are s spares. Thus, a failure will occur only if $(s+1)$ components fail. Any specific set of $(s+1)$ components will fail with a probability of $(1\text{-}a)^{s+1}$. The number of ways in which $(s+1)$ components out of n can fail is

$$f = \binom{n}{s+1} = \frac{n!}{(s+1)!(n-s-1)!} \qquad (2\text{-}10)$$

Actually, this relation and those of Equations (2-4) and (2-7) give the maximum number of failure modes. In actual practice, not all of these may bring down a processor. (In NonStop systems, for instance, a dual processor failure will only cause a system failure if a critical process pair is running in that processor pair). Thus, it is more accurate to express f as

$$f \leq \frac{n!}{(s+1)!(n-s-1)!} \qquad (2\text{-}11)$$

Following the previous examples, we have

$$F \approx f(1-a)^{s+1} \qquad (2\text{-}12)$$

$$A \approx 1 - f(1-a)^{s+1} \qquad (2\text{-}13)$$

where, from Equation (2-1),

$$(1-a) \approx \frac{mtr}{mtbf} \qquad (2\text{-}14)$$

Equations (2-11) through (2-14) along with Equations (1-5) and (2-1) are the fundamental relationships for analyzing the reliability of distributed systems such as active/active systems.

As an aside, note that many of these relationships are "approximately" equal. This is because we have only considered some of the failure modes. For instance, if we have a singly redundant system, we have considered only the failure of component pairs. The failure of three or more components will also take down the system.

However, the probability of this happening is so small that we can ignore it. This approximation error is analyzed in Volume 1 of this series,[30] where it is shown that the error induced by the approximation is less than 5% over the range of configurations in which we are interested.

The Importance of Repair Time

As we have discussed, there are two important factors that contribute to system reliability:

- its mean time before failure (MTBF) and
- its mean time to repair (MTR).

In a distributed network of computers such as an active/active system, the system MTBF and MTR are in turn determined by the component node (or subsystem) mtbf and mtr. Likewise, in multi-processor systems such as HP's NonStop servers, system MTBF and MTR are determined by the component processor (or subsystem) mtbf and mtr.

We usually cannot control the subsystem mtbf – presumably, the engineers have done all they can do to optimize that. But we can often influence the subsystem mtr. We can stock critical parts on site, we can purchase fast-response service level agreements (SLAs) or contract for on-site maintenance, we can take advantage of customer replaceable units, and so forth.

It turns out that reducing the subsystem mtr can have magnified effects on system availability and system MTBF. As we have

---

[30] See Appendix 2, Availability Approximation Analysis, *Breaking the Availability Barrier: Survivable Systems for Enterprise Computing*, AuthorHouse; December, 2004.

previously shown in Equations (2-13) and (2-14), system availability, A, is given by

$$A \approx 1 - f(1-a)^{s+1} \tag{2-13}$$

and subsystem availability, a, is given by

$$a \approx 1 - \frac{mtr}{mtbf}$$

or

$$(1-a) \approx \frac{mtr}{mtbf} \tag{2-14}$$

Substituting (2-14) into (2-13), we have

$$A = 1 - F \approx 1 - f\left(\frac{mtr}{mtbf}\right)^{s+1} \tag{2-15}$$

or

$$F \approx \frac{MTR}{MTBF} \approx f\left(\frac{mtr}{mtbf}\right)^{s+1} \tag{2-16}$$

where F is the probability of a system failure.

Thus, the probability of system failure is reduced by the square of subsystem mtr reductions for single-spared systems ($s=1$) and by their cube for dual-spared systems ($s=2$). If we can cut the subsystem repair time in half, we will reduce the probability of system failure by a factor of four if our system has one spare subsystem and by a factor of eight if it has two spare subsystems.[31]

---

[31] Why? When you reduce mtr, you reduce the length of downtime of the first component that failed. Hence, the probability of the failure of a second component during this shorter failure window is also reduced.

Sequential Repair

There are two types of repair that we will consider:

- *sequential repair*, in which multiple subsystem failures are repaired one at a time (for instance, there is available only one repairman)
- *parallel repair*, in which there is repair activity on each failed subsystem should multiple subsystems fail (for instance, there is one repairman for each subsystem)

We assume that repair times are random.[32] In order for the system to fail, there must have been $s+1$ subsystem failures, where s is the number of spare subsystems in the system. The system will be restored to service as soon as the first subsystem is repaired.[33] For sequential repair, this is the one subsystem on which repair efforts are underway. Thus,

$$MTR = mtr \quad \text{(sequential repair)} \quad (2\text{-}17)$$

From Equation (2-16), we see that the failure probability is proportional to $(mtr/mtbf)^{s+1}$. Thus, reducing mtr by a factor of k will decrease the probability of failure by k^{s+1}. From Equation (2-17), we

---

[32] The subsystem repair time is dependent on many factors – delay time waiting for a service technician, travel time to get to your site, analysis time to determine the fault, order time to get the component, another waiting time and travel time, doing the actual repair, and testing the result.

[33] Math nut note: Random repair times imply an exponential distribution, which is known as a memory-less distribution. As a consequence, should a node repair be in progress, its average repair time from that point is still mtr regardless of how long the repair has been in progress. See Chapter 4, Basic Performance Concepts, *Performance Analysis of Transaction Processing Systems*, by W. H. Highleyman, Prentice-Hall; 1989.

see that MTR is decreased by a factor of k. Therefore, MTBF will be increased by a factor of k^s.

This is the amplified impact of subsystem mtr. **In a singly spared system with sequential repair, if subsystem mtr is reduced by k, system MTBF will be increased by k; and the failure probability will be decreased by a factor of k^2. In a dually spared system, system MTBF will be increased by k^2; and the failure probability will be decreased by a factor of k^3.**

This leads to a restatement of Rule 8 given in Chapter 1 of Volume 1 of this series,

> **Rule 42**: *In a system with s spares, reducing subsystem mtr by a factor of k will reduce system MTR by a factor of k and will increase system MTBF by a factor of k^s, thus increasing system reliability by a factor of k^{s+1}.*

Parallel Repair

When a system with s spares fails, that means that there are $s+1$ subsystems in failure. Returning any one of these to service will restore the system.

Up until now, we have assumed sequential repair. The system MTR is the time that it takes to repair one of the subsystems. Thus, MTR = mtr.

However, what if there are several service people available, one per failed subsystem, and all subsystems undergo repair at the same time. (If the subsystems are separated by long distances, this would probably be the case.) Then the system will be restored when the first subsystem becomes operational.

Let us assume that the subsystem repair time, mtr, is random and that the repair time required for one subsystem is independent of the repairs on other subsystems. The repair rate for a single subsystem is 1/mtr (that is, given several subsystem failures, a service technician could repair subsystems at a rate of 1/mtr subsystems per unit time). For instance, if mtr were four hours, a technician could repair two subsystems in an eight-hour day.

If $s+1$ subsystems are down, and if $s+1$ service technicians are working on these subsystems, one per subsystem, the repair rate will be $(s+1)$ times faster, or $(s+1)$/mtr. Thus, the interval between repairs is mtr/$(s+1)$. Since MTR is the time to the first repair, then

$$MTR = \frac{mtr}{s+1} \qquad (2\text{-}18)$$

and the system MTR has been reduced by a factor of $(s+1)$. This leads to a generalization of our Rule 7, given in Chapter 1 of Volume 1:

Rule 43: *If $s+1$ subsystems fail and are being repaired simultaneously, and if the return to service of any one of these subsystems will return the system to service, the system MTR is mtr/$(s+1)$.*

The Impact of Parallel Repair

The basic failure probability equation that we have been using for a single-spared system is, from Equations (2-4) and (2-5),

$$F = \frac{n(n-1)}{2}(1-a)^2 \quad \text{(parallel repair)} \qquad (2\text{-}19)$$

Though not specifically mentioned before, this relationship applies to parallel repair. Keep in mind that the term $n(n-1)/2$ is the number of failure modes. In effect, there are $n(n-1)/2$ ways to pick two subsystems out of *n* to fail. If subsystems 3 and 4 were to fail, it would make no difference in what order they failed since both would undergo repair at the same time. The system is returned to service as soon as one of the two is repaired.

However, with sequential repair, the order is important since *only the first subsystem to fail will be repaired prior to returning the system to service*. Therefore, there are *n(n-1)* ways that the system can fail – one of *n* subsystems fails (which is the one to be repaired) followed by one of *n-1* subsystems. Thus, there are *n(n-1)* failure modes (*f*); and system availability is

$$F = n(n-1)(1-a)^2 \quad \text{(sequential repair)} \quad (2\text{-}20)$$

We have shown the impact of parallel repair on system MTR in Equation (2-18) (MTR = mtr/(*s*+1). We can now explore the impact of parallel repair on MTBF. For sequential repairs and using Equations (1-4) and (2-14), Equation (2-20) can be rewritten as

$$F \approx \frac{\text{MTR}}{\text{MTBF}} = n(n-1)\left(\frac{\text{mtr}}{\text{mtbf}}\right)^2 \quad \text{(sequential repair)}$$

If we move to parallel repair, we know that MTR will be cut in half in a single-spared system. However, MTBF will not be changed, since the repair time for the first subsystem to fail is still mtr. Thus, the failure window during which there can be a second failure is unchanged. Therefore, F is reduced by a factor of 2, which can be written as

$$F \approx \frac{\text{MTR}}{\text{MTBF}} = \frac{n(n-1)}{2}\left(\frac{\text{mtr}}{\text{mtbf}}\right)^2 \quad \text{(parallel repair)}$$

This is, of course, Equation (2-19). In effect, parallel repair has affected the number of failure modes, as noted above. Using this insight, the effect of parallel repair on multiply-spared systems can be deduced by looking at failure modes. To extend the argument above, the number of failure modes f for a sequentially-repaired system with s spare nodes is

$$f = n(n-1)(n-2)...(n-s) = \frac{n!}{(n-s-1)!} \quad \text{(sequential repair)}$$

We know from Equation (2-10) that the number of failure modes for a system with s spares and parallel repair is

$$f = \frac{n!}{(s+1)!(n-s-1)!} \quad \text{(parallel repair)}$$

Thus, we can conclude that *the number of failure modes are decreased by a factor of (s+1)! when parallel repair is used relative to sequential repair.* Since system failure probability is proportional to the number of failure modes [Equation (2-12)], the probability of system failure F will be reduced also by $(s+1)!$. But $(s+1)$ of this reduction is due to the reduction in MTR. Therefore, MTBF has been increased by a factor of $s!$ if parallel repair is used instead of sequential repair.

As examples, for a single-spared system, MTR is reduced by a factor of 2; and MTBF is unchanged, thereby reducing the probability of system failure F by a factor of 2. For a dual-spared system, MTR is reduced by a factor of 3, MTBF is increased by a factor of 2, and the probability of system failure is reduced by 6 [see Equation (2-7)].

This leads to Rule 44:

Rule 44: (Corollary to Rule 42) - *In a system with s spares, going to parallel repair will decrease MTR by a factor of (s+1), will increase MTBF by a factor of s!, and will decrease the probability of failure by a factor of (s+1)!.*

The Importance of Recovery Time and the 4 Rs

So far, we have assumed that a subsystem is returned to service as soon as it is repaired. This is a simplistic view if the subsystem is a node in an active/active network because once a node is repaired, it must be recovered.

Recovery of a node involves reloading the applications, resynchronizing its database, bringing up the network, and reconnecting the users, among other tasks. Until these tasks are completed, the system is short one node; and another node failure will take down the system if all spares are currently down.

Node recovery times can be significant. They can range from minutes to hours depending upon the complexity of the applications, the size of the database, the number of users, and, perhaps most importantly, the amount of time that the system has been down (the longer it is down, the more data must be recovered to synchronize its database).[34]

The Impact of Restore Time on MTR

Up until now, we have defined MTR to be the mean time to repair a system that has gone down (that is, it has suffered $s+1$ subsystem failures). The system is returned to service upon the repair of the first subsystem.

---

[34] An interesting approach still in research is the concept of microrebooting. Basically, if you can't eliminate all software errors, learn to live with them by trying to detect their potential occurrence and by rebooting that portion of code which is about to fail. See G. Candea, S. Kawamoto, Y. Fujiki, G. Friedman, A. Fox, Microrebooting – A Technique for Cheap Recovery, Proceedings of the 6th Symposium on Operating Systems Design and Implementation; December, 2004. See also Microrebooting for Fast Recovery, *Availability Digest*; March, 2007 (www.availabilitydigest.com).

However, once we introduce the concept of a recovery process following repair, MTR must be redefined. In fact, there are now four "r's" to consider as shown below and in Figure 2-2:

Repair   is the time required to repair a fault that took down a subsystem. Repair is generally the correction of a hardware fault.

Recovery   is the time required to recover a repaired subsystem so that it can be returned to service. It typically includes such tasks as loading the software environment and the applications, opening the databases, and verifying proper operation. The sum of the repair and recovery times is the time required to return the subsystem to service. It is the mtr for the subsystem.

Restore   is the time required to restore a failed system to service once the first node has been recovered. Restore tasks might include synchronizing the database of the failed node and reentering transactions that had been manually processed during the system outage. The sum of the subsystem mtr and the restore time is the time that it takes to return the system to service. This is the MTR for the system.

Return   is the time it takes to return a subsystem or the system to service once it has failed.

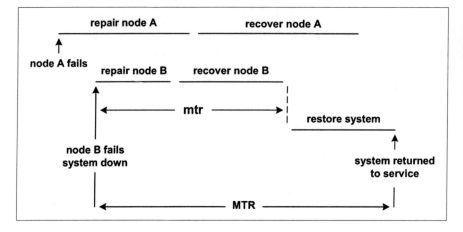

The 4 Rs
Figure 2-2

To be more precise in our "r's," we define the following terms:

r_h is subsystem *repair* time (typically hardware repair).
r' is subsystem *recovery* time following a repair, if any.
R is the system *restore* time following the return to service of one of the failed nodes.

Furthermore, we have now redefined the "r" in mtr and MTR to mean *return*:

mtr   = mean time to *return* a subsystem to service
       = subsystem repair time plus recovery time

MTR   = mean time to *return* a system to service
       = subsystem mtr plus system restore time

To ease the complexity of the following equations, let us represent the subsystem mtr by r:

r is the subsystem mtr. It is the mean time to *return* a subsystem to service.

Let us consider the case of parallel repairs. Following a dual failure of a singly redundant system ($s = 1$), subsystem mtr is r/2 [see Equation (2-18)], and[35]

$$MTR = r/2 + R \qquad (2\text{-}21)$$

That is, the time to return a system to service is the time that it takes for the first subsystem to be returned to service plus the time to restore the system.

In our previous analysis of a singly-redundant, fault-tolerant system, we found that its probability of failure was

$$F \approx \frac{n(n-1)}{2}(1-a)^2 \qquad (2\text{-}4, 2\text{-}5)$$

This analysis assumed that the system is returned to service as soon as a failed subsystem is returned to service; that is, MTR = $r/2$. However, we now know that MTR is further extended by the recovery time R as expressed in Equation (2-21). Since the system failure probability is proportional to MTR, our previously calculated failure probability must be increased by the factor

$$\frac{r/2 + R}{r/2}$$

and

$$F \approx \frac{r/2 + R}{r/2} \frac{n(n-1)}{2}(1-a)^2 \qquad (2\text{-}22)$$

---

[35] If we were to assume sequential repair, r/2 would be replaced by *r* in the following relationships.

This is the relationship incorporating system restore time for the reliability of a singly-redundant system with parallel repair.[36]

Impact of Hardware, Software, and Operator Faults

Up until now, we have assumed that all failures in a system comprising several like subsystems are the result of the failure of $s+1$ subsystems due to hardware faults. The return to service of the system requires that a subsystem be repaired and then recovered. However, in actuality, many subsystem failures may not be caused by hardware failures at all and will not require hardware repair. A software crash does not require hardware repair (though the software will probably be corrected at some later time). Nor does operator error require a hardware repair. These types of failures require only the recovery of the failed subsystem and the system restoration tasks to be performed.

If one looks at what happens following a subsystem failure that has taken down the system, there are three basic activities that take place *before* the system is ready for restoration. First of all, there is a decision time during which it is decided what to do to bring the subsystem back into service. Let us call this time d. Then there may be a repair time to correct a subsystem hardware fault. Let us call this r_h. Finally, at least one failed subsystem must be recovered and brought back on line. Let us call this time w. Further, let us call the subsystem return-to-service time for the case of no hardware failure r'. This is the decision time d plus the subsystem recovery time w. mtr continues to be defined as the total average time to return a subsystem to service.

> d is the average decision time following a system failure.
> r_h is the average time to repair a hardware fault.
> w is the subsystem recovery time.

---

[36] This relationship is formally derived in Appendix 3, <u>Failover Fault Models</u>, of Volume 1 of this series.

r' is the time to return a subsystem to service exclusive of hardware repair time. It includes decision time, d, and recovery time, w.

mtr is the mean time to repair a subsystem and recover it so that it can be returned to service.

Based on the above definitions, r' is the decision time plus the subsystem recovery time:

$$r' = d + w$$

Let us also define the probability h that a subsystem failure is due to a hardware fault:

h is the probability that a subsystem failure is due to hardware.

All subsystem repairs require a time of r'. In addition, h of these require a hardware repair time r_h. Thus, subsystem mtr, r, is

$$r = mtr = r' + hr_h$$

This value of mtr now can be used to obtain the failure probability for a system in which the subsystems comprising that system might be taken down by hardware faults, software faults, or operator errors. This easily can be done via Equation (2-22) by substituting $(r'+hr_h)$ for r. That is, the subsystem mean time to repair, mtr, is now $(r'+hr_h)$. Doing so results in[37]

---

[37] A more accurate analysis using state diagrams shows that F is a more complex function of the different nodal hardware and software availabilities. It can be shown that the simplified relationship of Equation (2-23), which depends only upon the overall nodal availability, a, is valid to the extent that $hr_h \gg d$ and $r_h \gg R$. For long hardware repair times, this tends to be true. Contact the authors for further information.

$$F \approx \frac{(r'+hr_h)/2+R}{(r'+hr_h)/2} \cdot \frac{n(n-1)}{2}(1-a)^2 \qquad (2\text{-}23)$$

Note the differences between r_h, r', and R:

> r_h is the hardware repair time.
>
> r' is the subsystem recovery time once the hardware has been repaired. It may involve deciding upon the action to take, reloading applications, opening databases, and so on. Only when the first failed subsystem has been recovered can system restoration begin.
>
> R is the system restore time once a subsystem has been returned to service. It may involve synchronizing the subsystem's database, entering backlogged transactions, and so on.

Assuming parallel repair of a single-spared system, the system MTR is half the subsystem mtr plus the system recovery time. Thus,

$$MTR = (r'+hr_h)/2+R \qquad (2\text{-}24a)$$

Also, since the probability of subsystem failure is $(1\text{-}a) \approx$ mtr/mtbf, the subsystem failure probability becomes

$$(1-a) \approx (r'+hr_h)/mtbf \qquad (2\text{-}24b)$$

where mtbf is the mean time between subsystem failures.

Equations (2-23) and (2-24) represent the failure probabilities and the return-to-service time for the case of a singly-spared system made up of multiple subsystems subject to parallel repair. A subsystem can fail due to a hardware fault requiring repair or a software fault or operator error not requiring repair. These equations are equivalent to

Equations (2-21) and (2-22) with mtr (i.e., *r*) expanded to show the effect of partial hardware faults.

What is r' and What is R?

As defined above, *r'* is subsystem recovery time exclusive of hardware repair and *R* is system restore time. But how are recovery tasks allocated to a subsystem or to a system?

Equation (2-23) holds the answer. If a task is part of the parallel repair (that is, each repair person will be doing that task), then it is a subsystem recovery activity. That is because it contributes $1/(s+1)$ of its time to MTR as shown in Equation (2-23) (which is for a single spare, or $s = 1$). Loading the applications and opening the databases are certainly subsystem recovery tasks. These will be done in parallel for all failed nodes if parallel repair is being done.

However, if a task cannot be performed until after the first subsystem is returned to service, this task is a system restore task. Its entire time will be added to the system MTR as shown in Equation (2-23). Database resynchronization is an example of a task that may either be done by a repair person in parallel with other subsystem repair efforts and therefore is a subsystem recovery task, or it may be done only after any one of the failed subsystems is available and therefore is a system restore task. Holding up putting the system back into service until the backlog of transactions that had accumulated during the failure are entered into the system is certainly a system restore task.

Note that if sequential repair is being performed, it does not make any difference mathematically which tasks are deemed to be subsystem recovery tasks and which are deemed to be system restore tasks. This is because MTR becomes

$$MTR = (r'+hr_h)+R$$

63

Both subsystem recovery time r' and system recovery time R are directly additive.

Nodes and Systems

So far, we have talked about systems and subsystems. However, in the active/active systems in which we are interested, there are really two levels of systems made up of like subsystems. At one level are the nodes which may comprise multiple processors. At a higher level is the system which comprises multiple nodes.

As an example, let us look at a node made up of sixteen processors (such as an HP NonStop server). Assume that following a node failure, the average decision time plus processor reload time r' is two hours, that processor hardware repair time r_h is 24 hours, that the average system restore time R is four hours, and that 20% of all failures are caused by hardware failures ($h = .2$). Furthermore, let the processor mtbf be 10,000 hours:

$$
\begin{aligned}
n &= 16 \text{ processors} \\
r' &= 2 \text{ hours} \\
r_h &= 24 \text{ hours} \\
R &= 4 \text{ hours} \\
h &= 0.2 \\
mtbf &= 10,000 \text{ hours}
\end{aligned}
$$

There is a single spare, and parallel repair is used. The processor (subsystem) mtr, the processor probability of failure (1-a), the node (system) MTR, the node probability of failure F, the node MTBF, and the node availability A for this case are shown in Table 2-2.

$mtr = r' + hr_h = 6.8$ hours

$(1-a) = mtr/mtbf = (r'+hr_h)/mtbf = 6.8x10^{-4}$

$MTR = (r'+hr_h)/2 + R = (2+0.2x24)/2 + 4 = 7.4$ hours

$F = \dfrac{(r'+hr_h)/2+R}{(r'+hr_h)/2}\dfrac{n(n-1)}{2}(1-a)^2 = \dfrac{7.4}{3.4}120(6.8)^2 x10^{-8} = 1.2x10^{-4}$

$MTBF = MTR/F = 7.4/1.2x10^{-4} = 62,000$ hours (7.1 years)

$A = 1-F = .99988 \approx$ four 9s

Node Availability
Table 2-2

However, when we consider an active/active system, the subsystem is now a node rather than a processor. The subsystem mtr is the MTR of the node, and the subsystem mtbf is the MTBF of the node. It is assumed that as soon as a node is returned to service following its repair and recovery, the active/active system is returned to service. There is no system hardware repair required ($h = 0$), nor is there a system restore time required ($R=0$). Let us take the case of a four-node active/active system with one spare. In this case,

n	= 4 nodes
mtr	= node MTR = 7.4 hours
mtbf	= node MTBF = $6.2x10^4$ hours

65

The subsystem (node) mtr, the subsystem probability of failure (1-a), the system MTR, the system probability of failure F, the system MTBF, and the system availability A are shown in Table 2-3.

mtr = 7.4 hours

$(1-a) = \text{mtr}/\text{mtbf} = 7.4/62{,}000 = 1.2 \times 10^{-4}$

$\text{MTR} = \text{mtr} = 7.4$ hours

$F = \dfrac{n(n-1)}{2}(1-a)^2 = 6(1.2)^2 \times 10^{-8} = 9 \times 10^{-8}$

$\text{MTBF} = \text{MTR}/F = 7.4/(9 \times 10^{-8}) = 8.2 \times 10^7$ hours (94 centuries)

$A = 1 - F = .99999991 \approx$ seven 9s

System Availability
Table 2-3

When Hardware Reliability No Longer Matters

As the hardware in a node becomes more and more reliable, there comes a point that further reductions in the hardware failure rate no longer have a significant effect on system availability. Where is that point?

In a previous section, we expressed in Equation (2-23) the failure probability of a single-spared system as a function of the probability, h, that a system failure is caused by a hardware failure:

$$F \approx \frac{(r'+hr_h)/2+R}{(r'+hr_h)/2} \frac{n(n-1)}{2}(1-a)^2 \qquad (2\text{-}23)$$

where

F is the probability of system failure.

r_h is the hardware repair time.

r' is the time to return a repaired node to service exclusive of hardware repair – the node's recovery time.

R is the system recovery time once a node has been returned to service.

h is the probability that a node failure is caused by a hardware failure.

n is the number of nodes in the system.

a is the availability of a node.

Since $(1-a)$ is equal to mtr/mtbf, where mtr and mtbf are the mean time to return the node to service and the mean time before failure of the node, respectively, and since the nodal mtr is $(r' + hr_h)$, Equation (23) can be rewritten as

$$F \approx \frac{(r'+hr_h)/2+R}{(r'+hr_h)/2} \frac{n(n-1)}{2}\left(\frac{r'+hr_h}{mtbf}\right)^2 \qquad (2\text{-}25)$$

The term mtbf is a function of a node's hardware mtbf, $mtbf_h$, and the software mtbf, $mtbf_s$. Since the rate of failure is 1/mtbf, the total node rate of failure is the sum of the hardware and software rates of failure:

$$\frac{1}{mtbf} = \frac{1}{mtbf_h} + \frac{1}{mtbf_s}$$

Therefore,

$$mtbf = \frac{mtbf_h \, mtbf_s}{mtbf_h + mtbf_s} \qquad (2\text{-}26)$$

67

Furthermore, the probability of a hardware failure can be expressed as

$$h = \frac{\text{rate of hardware failures}}{\text{total rate of failures}} = \frac{mtbf}{mtbf_h} \qquad (2\text{-}27)$$

The node mtbf and its hardware $mtbf_h$ can be measured over a period of time. The software $mtbf_s$ can also be measured, or it can be deduced from Equation (2-26).

If new hardware for the nodes is being considered, a reasonable question to ask is what impact will the availability of the new hardware have on system availability. If it is more reliable than the current hardware, will it increase the availability of the system significantly?

We assume that we know the hardware $mtbf_h$ of the new nodal system and that its other parameters stay the same (its repair time r_h, its nodal recovery time r', its system recovery time R, and its software mtbf, $mtbf_s$). The new nodal mtbf can be determined from Equation (2-26) and its probability of hardware failure from Equation (2-27). Using these values, the probability of failure for the new system can be determined from Equation (2-25).

There comes a point that increasing nodal hardware reliability has little effect on system availability because recovery and repair times become dominant. As hardware $mtbf_h$ grows larger, the probability of hardware failure, h, approaches zero; and the system mtbf approaches $mtbf_s$. In the limit, the system failure probability, F, becomes

$$F \approx \frac{r'/2 + R}{r'/2} \; \frac{n(n-1)}{2} \left(\frac{r'}{mtbf_s} \right)^2 \qquad (2\text{-}28)$$

Improving hardware reliability cannot improve system reliability beyond this point since node availability is controlled by node and system recovery time, not by hardware repair time.

This can be illustrated with the following example. Let

r_h	24 hours	(node repair time)
r'	2 hours	(node recovery time)
R	4 hours	(system recovery time)
n	2	(number of nodes)
$mtbf_s$	10,000	(software mtbf)

From Equation (2-28), the failure probability limit if hardware were not a factor is .00000012, or 6.9 9s. The system availability as it approaches this limit as a function of increasing hardware availability is shown in Figure 2-3.

The above analysis emphasizes the importance of minimizing recovery and restore times and validates Rule 21 given in Volume 1:

Rule 21: *System outages are predominantly caused by human and software errors.*

System Splitting

One method of designing an active/active system involves rearchitecting a monolithic system into a set of smaller, cooperating nodes, where the sum capacity of the nodes equals that of the original monolithic system plus perhaps some spare nodes. We call this *system splitting*. Let us explore this further for systems with no spares.

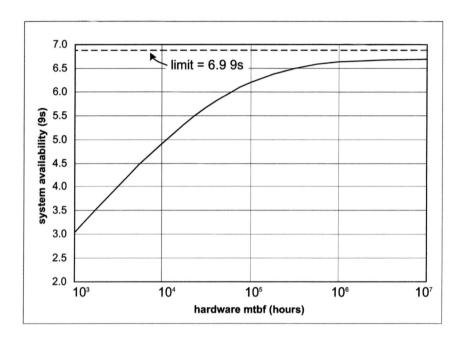

Impact of Hardware mtbf on System Availability
Figure 2-3

Multiprocessor Systems

We first analyze the impact of system splitting on multiprocessor, fault-tolerant systems such as NonStop servers. Let us take as a straw man a monolithic 16-processor NonStop system and break it into four four-processor nodes in an active/active architecture, as shown in Figure 2-4. Both systems have the same capacity, but they have markedly different reliability characteristics.

The monolithic sixteen processor system has a total of 120 failure modes [16x15/2 – see Equation (2-4)]. However, each of the nodes has only six failure modes (4x3/2). If we consider the system to be down if we cannot provide 100% capacity, then the active/active system will fail if any one node fails. Thus, it has 4x6=24 failure modes. This is one fifth the number of failure modes as the monolithic system. If the monolithic system has an MTBF of five years, the active/active system will have an MTBF of 25 years. This is an added advantage of active/active systems that we have not yet considered.

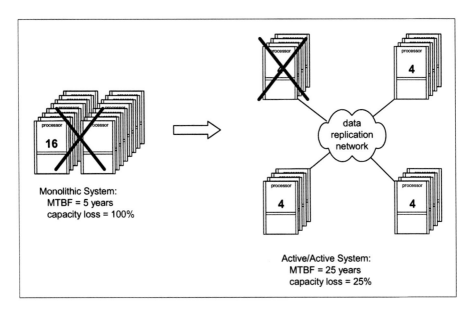

System Splitting
Figure 2-4

This concept can be generalized as follows. A monolithic NonStop system of n processors will fail with a probability, F_m, of

$$F_m \approx \frac{n(n-1)}{2}(1-a)^2$$

If the system is split into k nodes, then a node will fail with a probability, F_n, of

71

$$F_n \approx \frac{\frac{n}{k}\left(\frac{n}{k} - 1\right)}{2}(1-a)^2$$

There are k ways in which a node failure will take down the split system. Therefore, the probability of failure of the split system, F_s, is

$$F_s \approx k\frac{\frac{n}{k}\left(\frac{n}{k} - 1\right)}{2}(1-a)^2$$

Comparing the failure probabilities of the monolithic system to the split system, we have

$$\frac{F_m}{F_s} = \frac{\frac{n(n-1)}{2}(1-a)^2}{k\frac{\frac{n}{k}\left(\frac{n}{k} - 1\right)}{2}(1-a)^2}$$

or

$$\frac{F_m}{F_s} = k\frac{n-1}{n-k} > k \qquad\qquad (2\text{-}29)$$

The inequality is true because (n-1) is always greater than (n-k) for $k > 1$.

Thus, splitting a NonStop system into k nodes will increase its reliability by at least k. In the above case, for k = 4, the improvement is 4(15)/12 = 5, as we have previously determined. This result is expressed in Rule 9 given in Volume 1:

Rule 9: *If a system is split into k parts, the resulting system network will be more than k times as reliable as the original system and still will deliver (k-1)/k of the system capacity in the event of an outage.*

It is true that the loss of a node deprives the users of some capacity. However, if we have the choice of a monolithic system that will lose 100% capacity every 5 years or an equivalent active/active system that will lose 25% capacity every 25 years, the latter is certainly attractive. Furthermore, the probability that we will lose more than 25% capacity in this example, i.e., a second node failure, is almost never.

Thus, active/active NonStop systems provide significant reliability advantages even without spare nodes. Considering that a node is a component of the system in this context, and remembering that a node's failure rate, (1-*a*), is related to a node's mtbf and mtr by

$$\text{mtbf} \approx \frac{\text{mtr}}{(1-a)}$$

we have accomplished this reliability increase by reducing the failure probability (1-*a*) of the smaller NonStop node, thus increasing its mtbf. We assume that the node recovery time, mtr, is unaffected by node size and remains constant.

The above analysis has assumed that the failure probability of a NonStop node increases as the square of its size [n(n-1)/2]. This is certainly true for hardware failures in single-spared nodes such as NonStop servers, since it takes the failure of two components (two processors in this case) to take down the node. However, one can question to what extent the availability of a node really is affected by reduced node size. To the extent that this effect is linear or non-existent, the above relationship is compromised.

73

There are many reasons why a node will fail, and some of these may not be related to node size. To pursue this, let us look at the impact of node size on the most common causes of node failure – hardware failures, software faults, operator errors, failover faults, and environmental faults.

Hardware Failures are proportional to component count and component stress (for instance, the speed at which a chip is run). Smaller nodes have less components and may be run at slower clock speeds and lower temperatures. Therefore, they will have proportionally fewer failures. Hardware count is particularly pertinent in NonStop servers since system capacity is directly related to the number of processors provided. Since it would take two processor failures due to hardware faults to take a node down, failure probability due to hardware failures is proportional to the square of the node size.

Software Faults in a mature system running trusted software are caused by obscure sequences of events or by race conditions. In either case, they are generally a function of transaction rate. The higher the transaction rate, the more likely that a fault will be found. In an active/active system, each node carries only a portion of the transaction load; so we would expect that node failures due to software faults would be proportionally decreased. Most software faults in a NonStop server will only take down one processor. Therefore, failure probability due to most software failures is proportional to the square of the node size. However, some software problems could take down an entire node. In these cases, failure probability is directly proportional to the node size.

Operator Errors most often occur when the operator is trying to respond to a nonfatal failure and makes a fatal mistake. Pulling the wrong disk in a mirrored pair is a classic example of an operator fault, as is entering a wrong command. As Rule 19 in Chapter 5 of Volume 1 states, "When things go wrong, people get stupider." Since we expect smaller nodes to experience fewer hardware failures and software faults, we can expect that there will be a corresponding reduction in operator errors. Some operator errors will take down only one processor, leading to a failure probability proportional to the square of the node size. Others may take down an entire node, leading to a failure probability directly proportional to the node size.

There are, of course, some operational procedures (like backup), that must be done on each node and that are not a function of node size. Failure probabilities in these cases are independent of node size.

Failover Faults occur when a fault-tolerant node attempts to fail over to a spare component, and that failover fails. Again, the fewer the hardware and software failures, the fewer the failover faults. Since failover faults are triggered by a single component failure, failover faults are directly proportional to node size.

Environmental Faults such as power and air conditioning are typically not a function of node size. Failure rates for these systems will be substantially the same regardless of node size.

The system failure rate is the sum of the rates for each of the above failure mechanisms. We have already analyzed the failure rates for hardware and software faults and operator errors. We analyze failover and environmental faults in later sections.

Thus, it is reasonable to assume that in a NonStop active/active system, failure rates decrease with node size. To the extent that failure rates decrease proportionately less than the square of the node size, the system-splitting relationship of (2-29) becomes optimistic in that it predicts a greater than actual improvement, but it is still qualitatively valid and useful.

Uniprocessor Systems

One could apply the above reasoning to nonfault-tolerant nodes such as Unix systems and argue that the smaller a node, the fewer hardware and software errors and thus the fewer operator errors and failover faults. This argument will hold provided that each node is at least k times more reliable than the monolithic system, where k is the number of nodes into which the monolithic system has been split. This can be shown as follows.

Let the availability of the monolithic system be A. Thus, its failure rate is $(1-A)$. If we separate this system into k nodes, then each node will be x times more reliable than the monolithic system because each node is smaller. However, we consider that the system has failed if any node fails (remember, we are considering the case of zero spares). Since there are k ways that we can lose one node, then the failure rate of the split system will be

$$\text{split system failure rate} = k\frac{(1-A)}{x}$$

To achieve our goal, the failure rate must be less than that of the monolithic system:

$$k\frac{(1-A)}{x} < (1-A)$$

Thus,

$$x > k$$

where x is the required nodal reliability improvement factor.

This leads to

Rule 45: *If you break a monolithic system into k smaller nodes with no spares, the system will be more reliable than the original* *monolithic system provided that each node is more than k times as reliable as the monolithic system.*

Failover Time

What is Failover?

In a redundant system, when a component fails, the system invokes a backup component to take its place. This process is called failover – the functions of the failed component are failed over to the backup component.

Depending upon the system, failover can be measured in milliseconds (process pairs in a NonStop system), seconds (an active/active system), minutes (a cluster), or hours or more (a cold backup system). During the failover period, some system functions may not be available, or some users may not be provided service. Once the failover is complete, all services are restored, albeit at perhaps ·a reduced responsiveness if components are more heavily loaded.

The Impact of Failover Time on Availability

We start by assuming the worst case – that the system is totally unavailable during the failover time. This is the case for active/backup computer configurations as well as for clusters if a cluster node is providing unique services. We will cover the case of partial unavailability during failover later. Active/active systems are examples of this case since a failure of one node out of n denies service to only $1/n$ of the users.

When considering failover time, we have two sources of system failure:

- $s+1$ subsystems fail, where s is the number of spare subsystems, or
- one subsystem fails and the system must failover to a backup subsystem.

s+1 Subsystems Fail

We consider the case of parallel repair for a singly-spared system ($s = 1$). The contribution to the probability of system failure due to a multiple subsystem failure then is given by Equation (2-22):

$$F \approx \frac{r/2+R}{r/2} \frac{n(n-1)}{2}(1-a)^2 \qquad (2\text{-}22)$$

where

F is the probability of failure of the system.
r is the subsystem mtr (hardware repair time if any plus subsystem recovery time).
R is the system restore time.
n is the number of subsystems in the system.
a is the availability of a subsystem.

Failover

The system will be down for a failover time every time that there is a failure of a subsystem actively engaged in processing. This distinction is important. If a subsystem fails that is acting purely as a backup, then its failure will cause no downtime due to failover since no applications need to fail over.

Let m be the number of active subsystems in the system:

m = number of active subsystems in the system.

If each of the m subsystems has a mean time before failure of mtbf, then an m-active-subsystem system will have a single subsystem failure every mtbf/m units of time. Let us define MTFO as the mean time for failover to occur. Then the probability that the system will be down while it is failing over is

probability that system is down due to failover =

$$\frac{\text{MTFO}}{\text{mtbf/m+MTFO}} \qquad (2\text{-}30)$$

That is, once a failover has completed, there will be an average time of mtbf/m until the next subsystem failure occurs (remember that mtbf is defined as the mean time <u>before</u> failure, not the mean time <u>between</u> failures). At that time, a failover must take place, taking a time of MTFO. Therefore, the total time between subsystem failures is (mtbf/m + MTFO). The system is down MTFO of this time.

System Probability of Failure with Failover

The probability of system downtime during failover must be added to the system failure time due to a multiple subsystem failure. Taking the case of an n-node single-spared system with parallel repair,[38] the system failure probability becomes

$$F = \frac{r/2+R}{r/2}\frac{n(n-1)}{2}(1-a)^2 + \frac{\text{MTFO}}{\text{mtbf}/m+\text{MTFO}}$$
$$\approx \frac{r/2+R}{r/2}\frac{n(n-1)}{2}(1-a)^2 + \frac{\text{MTFO}}{\text{mtbf}/m} \qquad (2\text{-}31)$$

where F, r, R, n, m, and a are defined above and

---

[38] An example of a singly-spared n-node system is a cluster. For an active/backup system, n would typically be 2 unless there was one backup node for multiple systems. Active/active configurations are considered later.

mtbf   is the mean time before failure of a subsystem.
MTFO   is the mean time for failover.

The approximation is good if mtbf/m >> MTFO which will normally be the case.

The parameters for subsystem return-to-service time, r, subsystem mean time before failure, mtbf, and subsystem availability, a, are related by Equation (1-5) as follows:

$$mtbf = r/(1-a)$$

Thus, Equation (2-31) can be rewritten as

$$F \approx \frac{r/2+R}{r/2}\frac{n(n-1)}{2}(1-a)^2 + \frac{MTFO}{r}m(1-a) \qquad (2\text{-}32)$$

Note that, as a practical matter, MTFO should be no larger than R, the system restore time, since the time to bring up the backup system should be no longer than the time to bring up the entire system following a multiple subsystem failure. In fact, it will generally be much less (except perhaps for a cold standby).

Some Examples

A Cluster

Let us first take the case of a single-spared four-node cluster which is made up of nodes with availabilities of .999. If two nodes fail, the system is deemed to be down since it is supplying less than 100% capacity. This is the case of a dual node failure. All nodes are actively engaged in processing. The mean time to return a node to service, mtr, is two hours, as is the system restore time. The failover time is three minutes. Thus,

$$
\begin{aligned}
r & \quad = 2 \text{ hours} \\
R & \quad = 2 \text{ hours} \\
n = m & \quad = 4 \\
a & \quad = .999 \\
\text{MTFO} & \quad = .05 \text{ hours}
\end{aligned}
$$

The probabilities of failure for this case are:

Probability that the system is down due to a multiple node failure $= 1.8 \times 10^{-5}$
Probability that the system is down during failover $= 10 \times 10^{-5}$
Probability that the system is down $= 11.8 \times 10^{-5}$

The inherent system availability has been reduced from .999982 to .999882, or from a little less than five nines to a little less than four nines, due to failover times. Failover time has reduced availability by about one 9. Put another way, failover time has increased downtime by more than a factor of six.

Active/Backup

Let us now take a fault-tolerant system with an availability of four 9s backed up by a like system. Of the two nodes ($n = 2$), only one is actively processing ($m = 1$). The mean time for failover is 2 hours. All other parameters are the same:

$$
\begin{aligned}
r & \quad = 2 \text{ hours} \\
R & \quad = 2 \text{ hours} \\
n & \quad = 2 \\
m & \quad = 1 \\
a & \quad = .9999 \\
\text{MTFO} & \quad = 2 \text{ hours}
\end{aligned}
$$

The results for this case are

Probability that the system is down due to a multiple node failure $= 3 \times 10^{-8}$
Probability that the system is down during failover $= 10^{-4}$
Probability that the system is down $= 10^{-4}$

The inherent system availability has been reduced from over seven 9s to four 9s. Failover time dominates the system availability in this case.

Active/Active Systems

Active/active systems and systems with similar characteristics are somewhat different in that only a portion of users are affected by a failover. If users are evenly distributed across the n nodes in the system, then the failure of a node affects only $1/n$ of the users.

Furthermore, all nodes are active so that $n = m$.

In many applications, availability is taken as the availability of services to the user, not of the availability of the system as a whole. In this case, a failover affects only $1/n$ of the users, and therefore the probability that the system is unavailable due to failover should be reduced by that factor.

Equation (2-32) then becomes

$$F \approx \frac{r/2 + R\, n(n-1)}{r/2} \frac{}{2}(1-a)^2 + \frac{MTFO}{r} \frac{m}{n}(1-a)$$
$$= \frac{r/2 + R\, n(n-1)}{r/2} \frac{}{2}(1-a)^2 + \frac{MTFO}{r}(1-a) \quad \text{for } m=n$$

(2-33)

Let us take as an example an active/active system with the same parameters as the active/backup example above, except that the failover time is 1 second. Then

$$
\begin{aligned}
r &= 2 \text{ hours} \\
R &= 2 \text{ hours} \\
n &= 2 \\
a &= .9999 \\
\text{MTFO} &= 1 \text{ second} = .00028 \text{ hours}
\end{aligned}
$$

The results for this case are:

Probability that the system is down due to a multiple node failure $= 3 \times 10^{-8}$
Probability that the system is down during failover $= 1.4 \times 10^{-8}$
Probability that the system is down $= 4.4 \times 10^{-8}$

The probability of failure has been increased by about 50% due to failover time, but the attribute of extreme availability has been maintained (over seven 9s).

The above examples lead to the following:

Rule 46: *Don't underestimate your failover time. It may well be the most important factor in perceived availability.*

Failover Faults

Failover faults are another failure mechanism for redundant systems. A failover fault occurs when a hardware or software failure is detected, and the system attempts to failover to a backup component but is unsuccessful. As a result, the system fails.

Let us consider a singly redundant system that survives by switching over to a backup subsystem should a subsystem fail. Thus, there are three ways that this system can become unavailable to the users:

- Two subsystems can fail.
- The system may be in the midst of a failover.
- One subsystem can fail, and the failover mechanism fails.

As we have seen from Equation (2-32), repeated below, for the case of a singly redundant system with parallel repair, the probability that two subsystems will fail or that the system is in the middle of a failover is

Outage probability due to dual failures or failover $\approx$

$$F \approx \frac{r/2+R}{r/2} \frac{n(n-1)}{2}(1-a)^2 + \frac{MTFO}{r}m(1-a) \qquad (2\text{-}32)$$

Let us define the probability p of a failover fault as

$p =$ probability that a failover attempt will fail (a failover fault)

A failover fault can occur after any single subsystem failure. A subsystem failure occurs with a probability of $(1-a)$, and there are n ways in which a single subsystem can fail.

When a failover fault occurs, subsystems do not need to be recovered. It is necessary only to restore the system using the surviving subsystems. Thus, subsystem mtr, $r,$[39] is replaced with system restore time, R. Consequently, the probability of failure for this node is modified by R/r. The probability of a failover fault is the combination of the above parameters and is

---

[39] As defined earlier, mtr is the subsystem hardware repair time plus its recovery time.

outage probability due to a failover fault $\approx$

$$\frac{R}{r}pm(1-a)$$

where m is the number of active subsystems in the system.

Recognizing that a system outage will be caused by a dual failure or a failover only $(1-p)$ of the time, and assuming that $p \ll 1$, we have the probability that the system will be down, F:

$$F \approx \frac{r/2+R}{r/2}\frac{n(n-1)}{2}(1-a)^2 + \frac{MTFO}{r}m(1-a)+\frac{R}{r}pm(1-a) \quad (2\text{-}34)$$

In some cases, a failover fault will affect only the users at one active node (this is the case, for instance, in an active/active system).[40] As was done earlier for failover times, the failover term and the failover fault term are divided by m, the number of active subsystems, yielding

$$F \approx \frac{r/2+R}{r/2}\frac{n(n-1)}{2}(1-a)^2 + \frac{MTFO}{r}(1-a)+\frac{R}{r}p(1-a) \quad (2\text{-}35)$$

To obtain a feel for the effect of failover faults on system reliability, let us consider the following case:

r	= subsystem mtr = 4 hours
R	= system restore time = $r/2$ = 2 hours
n	= number of subsystems = 16
a	= subsystem availability = .999

---

[40] For instance, in an HP NonStop system, if a disk process fails and the failover is unsuccessful, the entire system is down. In an active/active system or in certain cluster configurations, if a node fails, only the users at that node are affected.

85

p = probability of a failover fault = .01
MTFO = mean time to failover = 0

Furthermore, assume in this case that all nodes are active ($n = m$) and that a failover fault will take down the entire system [that is, Equation (2-34 holds].

Using $R = r/2$ and $m = n$, Equation (2-34) is simplified to

$$F \approx n(n-1)(1-a)^2 + \frac{pn}{2}(1-a)$$

If there were no failover faults ($p=0$), then $F=.00024$ (an availability of .99976). However, with a failover fault probability of 1%, then $F=.00032$. A 1% chance of a failover fault makes the system 33% less reliable!

This effect is even more serious for a smaller number of nodes. Consider a system with the above characteristics except that there are only two nodes ($n = 2$). The above relationship then reduces to[41]

$$F \approx 2(1-a)^2 + p(1-a)$$

This can be rewritten as

$$F \approx [2(1-a)][(1-a)+p/2]$$

This can be quite easily interpreted. A first node will fail with a probability of $[2(1-a)]$. The second node will fail with a probability of $[(1-a) + p/2]$ rather than with a probability of $(1-a)$ as it would if there were no failover faults. The effective failover probability of the second node has been increased by $p/2$.

---

[41] Thanks to Alan Wood and Carl Niehaus of the NonStop Enterprise Division of Hewlett-Packard for this insight.

In this example, if there were no failover faults, the failure probability would be .000002. Given a 1% chance of a failover fault, the system failure probability is, in fact, .000012, or six times greater. A 1% probability of a failover fault makes the system 500% less reliable!

This effect is even more pronounced if the system comprises subsystems whose failure probability is much less than the failover fault probability. For instance, consider the case in which the subsystems have an availability of four 9s. Then $p \gg (1-a)$ [in this case, $p = .01$ and $(1-a) = .0001$], and the above expression can be written as

$$F \approx (1-a)p$$

Thus, in effect, if $p \gg (1 - a)$ (it is 100 times bigger in this example), once one system fails, the system acts as if the surviving system has a failure probability of p instead of $(1-a)$. Thus, in this example, the system availability is significantly less than we would expect since $p \gg (1-a)$.

Failover faults are indeed serious as expressed in the following rule:

Rule 47: *A small probability of a failover fault may cause a disproportionate decrease in system availability. Moving from active/backup to active/hot-standby or active/active with frequent testing can significantly reduce the probability of failover faults.*

Environmental Faults

There is one additional failure mechanism that must be recognized and accounted for, and that is an environmental fault. An environmental fault is any happenstance external to the computing system that will cause a system outage. Environmental faults can range from computer room fires or power outages to hurricanes and earthquakes.

From an availability calculation viewpoint, environmental faults are different from other faults which we have considered because they are asymmetric. If a node in a distributed system has a given mtbf and a given mtr, we often apply those parameters to all nodes in the network. However, environmental faults are site-dependent. A node in Florida is more likely to be taken out by a hurricane than is a node in Idaho.

An environmental fault is just another way for a node to fail, and so it affects the node's availability. Once we understand the availability of each node including environmental faults, all the rest of our analyses hold. Therefore, let us look at nodal availability in the presence of an environmental hazard.

We consider that a node can fail for one of two reasons – a system fault or an environmental fault. System faults include all of the failure causes that we have discussed so far – hardware faults, software faults, operator errors, and failover faults. Though we have previously included environmental faults in this list, we are now going to consider them separately.

As we have defined previously, let mtbf be the mean time before failure for a node due to system faults, and let mtr be the time required to return a node to service (including hardware repair and node recovery).

Environmental fault calculations are generally not reliable in that it is hard to state a probability that an earthquake will occur at a given site in the next ten years. However, one can take educated guesses which should err on the conservative side. For instance, one might reasonably assume that a data center in Florida will be hit every ten years by a hurricane. The Nile River floods every year for weeks. Power failures in third world countries could occur every day.

For a given site, one must estimate what the probability of an environmental fault is over a stated period of time. Alternatively, this can be stated as a mean time between environmental faults. If the mean time between hurricane disasters for a given site is estimated at ten years, then the probability that the site will be damaged by a hurricane in the next year is 10%.

Likewise, one must estimate what the mean time to restore for a damaged site will be. It is probably not the four hours that it would take to restore a system that has suffered a software fault. There may be a delay before the site can be accessed. A new system may have to be ordered and installed. A new facility may have to be acquired. The mean time to restore the services provided by the site might be measured in days or weeks. During this time, the application network is running without a spare, and a failure of a backup system may cause a total system outage.

Let us define mtbe as the mean time between environmental faults for a site, and mtre as the mean time to restore the site following an environmental disaster:

mtbf =   mean time before node failure due to a system fault.
mtr =   mean time to return a node to service following a system fault.
mtbe =   mean time before an environmental fault.
mtre =   mean time to return a node to service following an

environmental fault.

The probability that the node will be down due to a system fault is mtr/(mtbf+mtr). The probability that the node will be down due to an environmental fault is mtre/(mtbe+mtre). Therefore, the probability, f, that the node will be down is the sum of these two probabilities:

$$f = (1-a) = \frac{mtr}{mtbf + mtr} + \frac{mtre}{mtbe + mtre}$$

$$\approx \frac{mtr}{mtbf} + \frac{mtre}{mtbe}$$

(2-36)

where

f is the probability of failure of the node.
a is the availability of the node.

and where the approximation holds providing that the recovery times are much less than the failure intervals.

Let us illustrate this with an example. Consider a high availability node with an availability of three 9s and an mtr of four hours. This implies that its mtbf is 4000 hours (about a half a year). Assume that the node is sited in an earthquake region in which it is estimated that an earthquake strong enough to do serious damage will occur about once every twenty years (say, 160,000 hours), and that it will take four days (say, 100 hours) to replace the node and restore it to service. In this case, the true availability of the node is

$$f = (1-a) \approx \frac{mtr}{mtbf} + \frac{mtre}{mtbe} = \frac{4}{4000} + \frac{100}{160,000} = .0016$$

$$a = .9984$$

The probability of failure has been increased by 60%, and the node availability has been reduced from three nines to somewhat less than three nines.

In the general case, individual node availabilities should be calculated for each node to take into account their respective exposure to environmental faults. (Actually, this same technique can be used if the nodes are heterogeneous with different availability characteristics.)

Given the corrected values for node availability, our previous analyses hold. More specifically, if we have a two node system with availabilities of a_1 and a_2, the probability of a two node failure with parallel repair is

$$F \approx \frac{r/2 + R}{r/2}(1 - a_1)(1 - a_2) \qquad (2\text{-}37)$$

r, the nodal mtr, is the mean time averaged over all nodes in the system to return a node to service. Since the nodes are not symmetric, r is the weighted average of the mtr's for each node.

Likewise, we can adjust the expressions for failover times and failover faults. In our previous analyses, we noted that one node must first fail in order to trigger a failover time or a failover fault. A node will fail with a probability of $(1-a)$. Since there are n identical nodes, the probability of a first failure is $n(1-a)$.

Taking the case of a two-node active/active system, if the nodes are not identical and have availabilities of a_1 and a_2, respectively, then the probability of a first node failure is $(1-a_1)+(1-a_2) = (2-a_1-a_2)$. The failure probability for an active/active system with parallel repair is then given by the appropriate modification to Equation (2-35):[42]

---

[42] This expression holds if the mtr's for the two nodes are approximately equal. Contact the authors for an analysis of the case in which this is not true.

$$F \approx \frac{r/2+R}{r/2}(1-a_1)(1-a_2) + \frac{MTFO}{r}(2-a_1-a_2) + \frac{R}{r}p(2-a_1-a_2)$$

This discussion leads to the following rule:

Rule 48: *Pick your node locations in an active/active system carefully to minimize the chance that environmental hazards will outweigh the availability of the nodes.*

What's Next?

In this chapter, we have reviewed *why* active/active systems can achieve extreme reliabilities. Given this incentive, however, there is a lot to consider when we think about *how* we can build these systems.

In the next chapter, we review the "hows" which were covered in Volume 1 of this series. They include the synchronization of database copies via data replication, recovery from various failure modes, and controlling the cost of database replicates.

Chapter 3 – An Active/Active Primer

"The person that knows how will always have a job. The person that knows why will always be his boss."
- Diane Ravitch

In the previous two chapters, we analyzed a concept that has the potential of revolutionizing computing architectures by providing extraordinary reliabilities at ordinary costs.

There are some classic parallels to systems with this extraordinary capability. When you pick up the telephone, you expect a dial tone. When you turn on a light, you expect electrical power.

Occasionally, a large group of users may be taken out of service as evidenced by telephone overloads during crisis times or by the two Northeastern United States blackouts in the last four decades. However, most outages are due to failures in the "final mile" – typically a tree limb falling on wires.

This extreme level of availability is achieved by having highly redundant networks that can route around failures. Active/active systems achieve their reliability by providing an equivalent flexibility; but they do so, of course, on a much smaller scale.

In the transaction-processing world, the classic approach to recovery from a processing system failure has been tape backup. However, recovery can be a very slow process and may take hours or days as the database is reconstructed. Besides, handling volumes of tape is an expensive and error-prone process. This has led to the use of active/standby system pairs in which the standby database is kept up-to-date via data replication. Such a configuration can reduce recovery time to minutes or hours but requires purchasing a second system whose capacity is not available for the application.

Active/active systems represent a major paradigm shift for high availability systems. No idle capacity need be purchased for backup purposes, and recovery from a failure can be completed in seconds.

If active/active systems represent such a powerful approach to system architecture, why are they not evident in all critical systems today? The answer is that there are, in fact, many active/active installations today experiencing almost no downtime. However, these are generally one-of-a-kind implementations that have been custom designed and built. Until recently, there have been no generic solutions to building active/active systems. However, generic solutions are now beginning to appear; and with these will come an increasing number of such systems.

In this chapter, we explore the challenges inherent in building an active/active system. We summarize the problems and today's solutions about which we talked in Volume 1 of this series, *Breaking the Availability Barrier: Survivable Systems for Enterprise Computing,* and in the first two chapters of this volume. The rest of this volume then describes additional problems, the solutions for which are also of great importance to active/active systems.

A General Solution

A general but simplistic solution for achieving extreme reliabilities in computing environments is shown in Figure 3-1. Users, servers, and databases are distributed throughout a redundant network. Distributed redundant copies of the database provide database resiliency in the face of database loss due to a failure or a disaster.

A user can be connected to any server in the network for the duration of a transaction. If a server fails, the user can be connected to another surviving server. Each completed transaction is applied to all database copies in the network.

However, this simplistic architecture suffers from a major flaw – network delays. Even at the speed of light, it takes 25 milliseconds for a signal to make the round trip between New York and Los Angeles. Signals, in fact, travel through copper or fiber channels; and their speed through these media is about half the speed of light (not counting delays through repeaters and switches). Thus, it takes at least 50 msec. for a signal to make a round trip between the U.S. coasts.

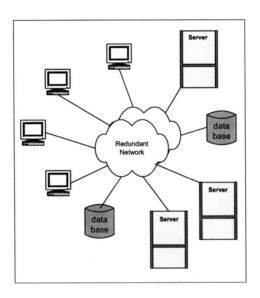

**General Active/Active System
Figure 3-1**

Consider the following. A user in one location is connected to a server in another location. This server must make updates to two or more databases at yet other locations. Each user/server interaction takes at least one round trip. Server updates to the database copies (assuming they are done in parallel) take one round trip for a start transaction, one or two round trips for each update (possibly a read followed by a write), and two round trips for the two-phase commit. Even a modest transaction with four updates will require up to twelve or more round trips.

95

Thus, network delays can add the better part of a second to transaction times. If subsecond response times are required, the capability to achieve them is seriously compromised by the architecture of Figure 3-1.

Network delays are a fundamental hurdle for active/active systems. The solution requires alternate architectures which themselves have issues. Let us look at these architectures, their challenges, and the solutions.

Database Locality

The fundamental problem of network delays is solved by *locality*. In normal operations, in order to avoid network delays, a copy of the database should be local to each server; and users should be local to their server.

However, should a database fail, servers using that database can look to another database in the network. Should a server fail, users can be switched to another server in the network. All users will continue to be provided with service, albeit at potentially reduced performance for some users.

An active/active system is then a network of independent peer nodes, each node serving its own local community of users. Each node is capable of operating on its own; but in normal operation, it must keep its peers informed of its own activity.

Database locality solves the fundamental network delay problem – there is nothing we can do about the speed of light – but raises a host of other issues:

- *Database Synchronization*: How do we keep the various database copies at each node synchronized so that every node has a current copy of the database?

- *User Access*: How do we ensure continued service to users serviced by a failed node?

- *Data Loss*: What do we do about database updates that may be lost if a node fails while it is distributing its database updates to other nodes?

- *Network Failure*: What do we do with a node that becomes isolated from the rest of the active/active system?

- *Referential Integrity*: Since each database is a working copy, how do we ensure its integrity in the presence of updates from many remote nodes?

- *Collisions*: What happens if users at different nodes update the same row at nearly the same time?

- *Deadlocks*: How do we resolve deadlocks across the network if we try to prevent collisions through locks?

- *Application Changes*: What is the extent of the changes required of the application to support an active/active architecture?

- *Performance*: What is the performance impact of maintaining synchronized databases across the network?

- *Security*: How do we protect the data that must be sent across the network to keep the databases synchronized?

- *Node Failure*: What impact will a node failure have on the rest of the system?

- *Node Recovery*: How do we reintroduce into the network a node that has failed and has been repaired?

- *System Expansion*: How do we add new nodes to the network?

- *Database Synchronization*: When a node is added to a system, how is the database of that node synchronized with the rest of the system?

- *Upgrades*: How do we upgrade hardware and software across the network?

- *Cost*: What are the real cost implications of active/active systems?

These issues represent those that must be overcome for a successful active/active implementation. The good news is that there are solutions today for each of these problems. Some of the solutions were presented in Volume 1 of this series and are summarized in this chapter. Others are introduced here and presented in later chapters.

Database Synchronization

Perhaps the most fundamental of the challenges mentioned above is the synchronization of databases across the network. This is best accomplished by data replication, a mechanism to propagate transactions between disparate databases.

Many data replication products exist today, and many of the problems listed above are addressed (or not addressed, as the case may be) by these products. Therefore, it is important to understand the nuances of data replication techniques. In this section, we will concern ourselves with asynchronous replication. Synchronous replication will be discussed in the next section.

The Data Replication Engine

A basic data replication architecture reflective of the preponderance of data replication engines today is shown in Figure 3-2. It depends on the availability of some sort of Change Queue, which contains the series of changes made to a source database. The Change Queue can take one of many forms:

- the audit trail, journal, or redo log created by a transaction manager such as Tuxedo, Oracle, HP's NonStop TMF, or IBM's IMS or CICS

- a log of changes created by the application or storage device drivers

- a log of changes created by database triggers.

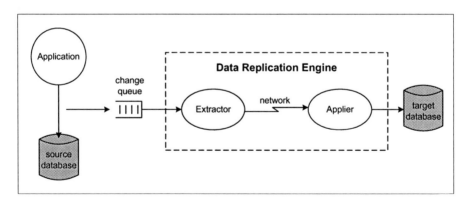

Basic Data Replication Engine
Figure 3-2

It is the job of the data replication engine to post these changes to a remote database. This is done via an Extractor/Applier pair. The Extractor extracts each change from the Change Queue and sends it over the network to the Applier, which applies that change to its target database.

Of course, for active/active systems, data replication must be bidirectional since each system is making changes to its local database and must replicate these changes to the remote database. This is generally done by providing two data replication engines, one replicating in each direction, as shown in Figure 3-3.

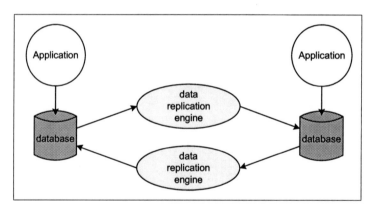

Bidirectional Replication
Figure 3-3

So far we have only talked about data replication between two nodes. However, an active/active system may have multiple nodes. Therefore, each node must have a data replication path to every other node in the network, as shown in Figure 3-4. Note that this data replication network need not be fully connected. Some replication paths may be routed through multiple nodes. In Figure 3-4, changes to node 1's database or to node 3's database are made to node 4 by picking up these changes from node 2. Likewise, changes to node 4's database are first made to node 2 and are then replicated to nodes 1 and 3. This represents a network cost/performance tradeoff – one example of our reliability/cost/performance compromises.

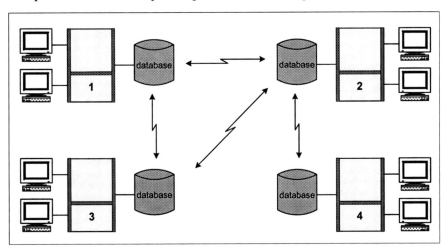

Multinode Replication
Figure 3-4

The data replication technique described above does not directly impact the application at all. Its activities occur unbeknownst to the application since replication proceeds independently of the application. Therefore, this type of replication is called *asynchronous* replication. Synchronous replication, its role, and its characteristics, are described a little later.

Currently available asynchronous replicators have several desirable characteristics:

- They may impose no response-time penalty on an application since replication is done "under the covers" with no involvement by the application.

- They are typically noninvasive in that there are no modifications that must be made to an application. Providing that there is a Change Queue that can be read by the Extractor, the multiple instances of the replication engine required to support the application network are simply configured.

- Depending upon their architecture (as discussed later), they can be very fast and can provide subsecond replication times.

- Their resource utilization footprint in terms of memory and required processor cycles can be quite efficient.

- A database copy need not be provided at each node so long as a node has access to a database across the network.

- Scalability is, to a first approximation, a function only of the number of database copies in the network and not the number of nodes since each node must only connect to those nodes containing databases.

- The network may be heterogeneous. Various nodes may use different hardware, different operating systems, different databases, or different versions of these.

However, there are some asynchronous replication issues that must be addressed. Though none of them are insolvable, they are serious and must be considered.

Ping-Ponging

Provision must be made to avoid ping-ponging. This is the reflection of an application's change back to the source database. When a change is made by an application to a database, that change is replicated to a target database by a replicator, where it is applied to the target. If the target Change Queue is an audit trail or is generated by triggers on the database, then this change will show up in the target's Change Queue. We do not want to rereplicate these changes back to the source database and set up an endless loop.

Therefore, bidirectional data replication engines must have provisions to mark or otherwise identify changes received from a remote node so that the return Extractors can ignore them.[43]

Data Loss

Because there is a propagation time (which we have called replication latency) to transmit a change from a source database to a target database, changes in the replication pipeline may be lost at the time of a failure of a node, of a database copy, or of the network.

Lost transactions, if any, must be recovered. A typical way to do this is to send a message to each user (or application process) who had been switched from a failed node to a surviving node and ask him to check whether his last transactions have been applied and to reenter them if they have not. This may leave a security hole, which must be addressed.

Collisions

It is quite possible (though not probable) that two users at different nodes will attempt to update the same row at nearly the same time – close enough that neither knows of the other's change, which has not yet found its way through the data replication engine. Both will replicate their change to the other system, and each replicated change will overwrite the original change in that system. As a result, the databases are now both different; and both are wrong. This is called a *data collision.*

---

[43] Strickler, G., et al.; *Bidirectional Database Replication Scheme for Controlling Ping-Ponging,* United States Patent 6,122,630; September, 19, 2000.

Data collisions must be detected and resolved, or they must be prevented. For instance, collisions may be detected by carrying with each change a version identification of the row that is being changed. If the version of the row to be changed is different from the row that has been changed, then a collision has occurred.

However, version checking may require a change to an existing database structure. The problem is avoided if the replication engine sends *before* images with the change data (the before image reflects the original value prior to the change). The before image which is sent must match the row to be updated in the target database, or else there has been a collision.

Once detected, collisions may be resolved in several ways. In some systems, collisions can be ignored. In others, they may be resolved by business rules. If all else fails, they must be resolved manually.

Later, we will describe a method for avoiding collisions altogether. This is by using synchronous replication, which, however, introduces a potential performance penalty. Nevertheless, if the occurrence of collisions requiring manual intervention is frequent enough, or if database discrepancy is prohibited by cost or regulation, synchronous replication may be attractive.

In addition, synchronous replication provides a consistent target database that is identical to the source database at any instant in time. Asynchronous replication provides a consistent target database, but one that lags the source database by the replication latency described next. This may be unacceptable in certain reporting applications or when fairness – the equal access to all data by all users – must be enforced.

Replication Latency

The delay that a change experiences from the time that it is made to the source database to the time that it is applied to a target database is called *replication latency.* We have already seen two examples of problems caused by replication latency – data loss and data collisions. The longer the latency, the higher the data loss at failover time and the higher the collision rate. Another problem is the return of stale data to a query.

The analysis of data replication performance is addressed in a later chapter. However, in summary, the amount of latency is affected by many factors. They include:

- disk queuing points
- polling delays
- communication buffering

Queues

As we have discussed, a data replication engine needs at least one persistent queue for recovery purposes – what we have called a Change Queue. It may use additional disk-resident intermediate queues for several reasons:

- to minimize the amount of lost data in the event of a network or node failure.

- to provide proper serialization of changes to the target database.

- to control the application of changes that are provided by nonrecoverable triggers at the source database (i.e., for backing out aborted transactions).

To the extent that there are additional queuing points, replication latency will be extended. Volume 1 of this series shows that queuing points can be avoided in many cases.

Polling

One technique for feeding components of a replication engine is for components to periodically poll their sources to see if there is any work to do. The polling period adds directly to the replication latency in addition to imposing additional processor load.

The most efficient replication engines are event-driven rather than poll-driven. An engine component is idle until a source has work for it to do. At that time, the component is immediately awakened and given the work.

Communication Buffering

A special case of queuing is batching changes to be sent over the communication channel connecting the Extractor to the Applier. Typically, rather than sending each change one at a time, changes are sent in batches whenever a full batch is accumulated or when a time limit has expired.

This buffering time adds directly to latency time. Its primary purpose is to make the use of the communication channel more efficient since each transmission incurs additional protocol overhead independent of the number of changes being sent. Communication buffering is a compromise between communication efficiency and replication latency.

Replication Capacity

Another performance issue with data replication is replication engine capacity. If nodal transaction rates are very high, a single-threaded replication engine may not be able to propagate changes fast enough. There may be multiple applications all funneling their transactions through a single replicator instance.

One possible solution is blocking. If changes can simply be read from the Change Queue in large blocks, sent to the Applier as large blocks, and perhaps in some systems even applied to the target database as large blocks,[44] then the change replication rate could be significantly faster than that achieved by reading changes one at a time from the Change Queue. Another solution might be to replicate bulk data operation procedures such as "delete rows where ..." via SQL statements or other operators.

Generally, these changes must still be written to the target database one at a time since the rows to which the changes apply are distributed randomly across the database. Since disk writes are often much slower than disk reads, the bottleneck created by the target database nullifies the effect of source blocking.

Furthermore, some forms of Change Queues do not lend themselves to blocked disk reads. Some audit trails, for instance, are pointer lists that point to changes stored in other tables. However, these changes can be blocked prior to transmission.

A viable method to increase replication capacity is to multithread the replication engine. There are fundamentally three components that can be multithreaded by creating multiple instances (or threads within

---

[44] E.g., some databases support row-set or other block-oriented operations.

an instance) of them – the Extractor, the communication channel, and the Applier.[45]

Figure 3-5 shows some different strategies for multithreading. In the most general case (Figure 3-5a), each of the component classes are independent of the others. A class of Extractors reads changes in parallel from the Change Queue and queues them for the communication channels. The communication channels read changes from their queue and propagate them to a queue feeding the class of Appliers. The Appliers read their queue and apply changes to the target database in parallel.

As a general statement, providing n threads will increase a replication engine's capacity by a factor approaching n. In this example, it is, of course, unnecessary to have the same number of instances for each component. One could, for example, provide fewer Extractors than Appliers if Change Queue reads were faster than target updates. The number of communication channels will depend on the speed of a channel. More WAN channels may be needed than LAN channels, for instance.

Though the strategy shown in Figure 3-5a introduces additional queues, these are typically memory-resident queues. The associated queuing delays can be made arbitrarily small by providing sufficient capacity via multithreading.

There are other multithreading strategies, as shown in Figure 3-5. For instance, each Extractor might have assigned to it its own communication channel, with the Extractor/channel pairs feeding a common queue on the target system for the Appliers (Figure 3-5b).

---

[45] Referential integrity issues associated with multithreading are discussed in detail in Volume 1, Chapter 10, Referential Integrity, and are reviewed here.

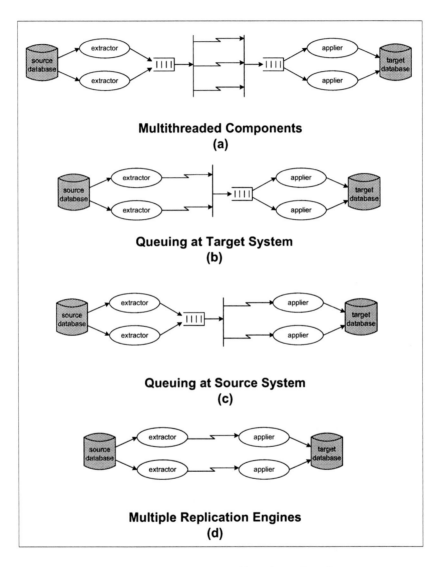

Multithreaded Components
(a)

Queuing at Target System
(b)

Queuing at Source System
(c)

Multiple Replication Engines
(d)

Multithreaded Replication Engines
Figure 3-5

Alternatively, the Extractors could feed a queue on the source system from which channel/Applier pairs pick up changes to be applied to the target database (Figure 5c). Finally, independent Extractor/channel/Applier threads could be used (Figure 5d). This last example is, in effect, the use of parallel data replication engines.

The increased capacity provided by multithreading, though not directly impacting replication latency, can have a positive secondary effect on latency. To the extent that the replication engine depends upon other disk queuing points, these queues will drain faster, thus reducing queuing delays. Communication buffers may fill faster, reducing communication buffering delays. Polling delays will be reduced because there are multiple threads polling a queue for activity.

The capacity increase afforded by multithreading and its impact on replication latency is analyzed later in Chapter 10 in Volume 3, which deals with replication engine performance.

Referential Integrity

Database Consistency

A fundamental problem introduced by multithreading is the maintenance of database consistency. The problem is caused by the replication of changes in parallel. If changes are all sent in order over a single replication channel, they are guaranteed to be applied to the target database in the same order that they were applied to the source. However, this order cannot be guaranteed if several independent replication threads are used. As a consequence, changes made to the source database may be applied in some unorganized order to the target database.

Keep in mind that a fundamental precept of active/active systems is that all database copies are being actively used by some instance of the application. Though these databases may be slightly different at any point in time due to replication latency, they must always be consistent. Providing updates to the target in some order that is different from the way that they were applied to the source may destroy the consistency of the target databases.

Natural Flow

A single-threaded replication engine guarantees that updates made to the target database can be done in the same order as they were done at the source system.[46] Thus, the *natural flow* of the change stream is maintained. If natural flow of updates is enforced at the target database, then the target database is guaranteed to be always in a consistent state provided that the source database is also always in a consistent state. In fact, the target database will always be an exact copy of the source database but delayed in time by the replication latency.

With a multithreaded replication engine, one change may flow along a different path than its preceding and following changes. The transit time of each path may be somewhat different and may result in a reordering of these changes as they are received at the target system. If the changes are applied to the target database as they are received, then natural flow has been violated.[47]

This can cause a host of problems relative to data consistency:

---

[46] Some asynchronous replication engines queue the changes and wait for the commit before applying the changes to the target database. These engines violate strict natural flow.

[47] This is true for related changes. Unrelated changes may be sent over other threads since the relative order of applying these changes is not important.

- If a row is inserted into a database, any other rows which it references must already exist. Typically, a child row will reference its parent. Therefore, if a child is inserted prior to its parent, an unresolved reference is created. This will ultimately correct itself when the parent row is received unless the target database's referential integrity feature is turned on at the row or event level. In this case, the child insert will be rejected. The condition is generally not recoverable.

- The same condition exists if an attempt is made to delete a parent while its child rows still exist. In general, a child cannot be inserted until its parent exists; and a parent cannot be deleted until all of its children have been deleted.

- If two consecutive updates change the value of the same row and are applied out of order, the row will be left with the wrong value since the older change will have overwritten the newer change.

- Improperly ordered changes may create other problems if the database is making referential integrity checks. For instance, if a credit to an account is followed by a debit to that account, and if the debit is applied before the credit at the target database, then this may result in a negative balance, which may cause the debit to be rejected at the target database.

To make matters even worse, there are " physical" or "base-level" replication engines that do not replicate logical updates but rather propagate physical changes to the database. Thus, data changes are replicated separately from index changes, allowing index values that do not yet correspond to an actual data item to be stored in the target database.

The violation of any consistency constraint such as those described above violate the *referential integrity* of the database. Some violations are temporary in nature and are resolved as replication continues. In this case, an application reading the database may get bad data, but it is a temporary condition. In other cases, such as the rejection of an insert, permanent damage may be done to a database.

This discussion leads to the following rule:

Rule 47: *Replication engines that violate referential integrity on the target database are rarely, if ever, suitable for active/active implementations.*

Serialization

It is of paramount importance that the referential integrity of all database copies in an active/active system be maintained. Therefore, changes must be applied to the target database in natural flow order. In a multithreaded replication engine, this means that a *serializing* facility that will restore natural flow for all related database changes is needed following all data replication threads and before the target database.

There are actually two levels of natural flow in a transaction-oriented system. The system could be designed for total natural flow of all events, or it could be designed for just the natural flow of transactions. If intratransaction referential integrity is not being checked at the target database, then natural flow of changes within a transaction can be violated (unless the same data item is updated multiple times within a single transaction) since no application can see these changes anyway until the transaction commits (changed records are locked during a transaction). However, if changes are checked during a transaction, then related changes must be applied in natural flow order. In any event, intertransaction natural flow must be observed.

If all events are to be applied in natural flow order, a performance bottleneck may be created since events must be written to the target database one at a time. Guaranteeing just the natural order of transactions can solve this problem since intratransaction updates can be applied in parallel so long as the order of transaction commits are maintained. In addition, nonrelated transactions can be committed in parallel.

Chapter 10, <u>Referential Integrity</u>, of Volume 1 of this series, delves deeply into serialization strategies for multithreaded replication engines. These techniques can be summarized in a few examples, as shown in Figure 3-6.

The simplest method is to use independent threads, as shown in Figure 3-6a. If update activity can be separated into isolated sets of tables, where all tables in a particular set are related to each other but are not related to tables in any other set, then all updates to a table set can be sent over a thread dedicated to that table set (or sets). Since these changes are independent of changes sent over other threads, then they can be applied independently without concern for interthread database consistency, even though natural flow is not strictly followed.

One problem is that a transaction may involve changes to multiple table sets and therefore will use multiple threads. A means must be implemented to ensure that all threads have completed changes for that transaction before it is committed.

If separate table sets cannot be defined, or if changes made are predominantly to one such set, then changes must be sent in parallel over multiple threads and reserialized prior to applying them to the target database. One way to do this is to write the multiple change streams to a Database of Change (DOC) at the target system (Figure 3-6b). In effect, the DOC is the recreation of the Change Log at the reserialization point. A Reserializer can then read changes from the DOC and send separate transactions to separate Appliers.

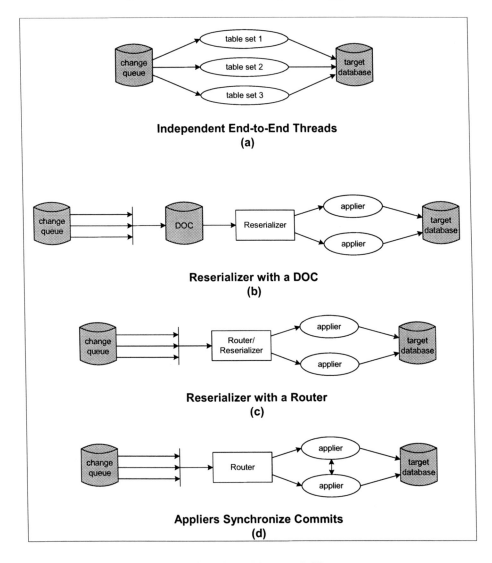

Enforcing Natural Flow
Figure 3-6

The DOC is useful in order to avoid posting aborted source transactions to the target and then having to back them out. This is particularly important if before images of data are not replicated. A DOC can also provide some fault-recovery benefits by receiving and storing changes to the target database when that database is offline for maintenance.

However, the DOC represents a disk-queuing point and will extend replication latency. An alternative is simply to queue changes in memory to a Router, which will route all of the changes for a given transaction to a specific Applier, as shown in Figure 3-6c.

In the technique of Figure 3-6c, intratransaction natural flow is guaranteed if all changes for a transaction are passed to a single Applier. However, additional measures are required to ensure the more important natural flow of transactions. Some typical procedures for this include:

- *Pause on Commit*: The Reserializer will pass changes for a transaction to a particular Applier, which will proceed to apply changes as they are made. However, when it is time to commit one of the transactions, the Reserializer will delay that commit until all previous transactions have been committed. It will then commit the delayed transaction.

  There is a chance for deadlock in this strategy if the changes received over independent threads are not first serialized. Should a later transaction lock a row before an earlier transaction attempts to lock it, then the earlier transaction cannot complete. This also means that the later transaction's commit will never be issued as it is awaiting the completion of the earlier transaction. In this case, the Appliers can coordinate between themselves; and the Applier with the later commit will release its locks.

116

- *Commit Reserialization*: To avoid holding up the change stream when a commit is underway, the Reserializer can send transaction changes to Appliers as above, including all commits. The Appliers then coordinate among themselves to ensure proper commit order, as shown in Figure 3-6d.

The requirement for proper serialization means that replication runs at the speed of the slowest thread, but this still represents a significant capacity improvement when compared to a single-threaded replicator.

Synchronous Replication

Two serious issues associated with asynchronous replication are data loss following a failure and data collisions. Both of these increase with replication latency, and both can be corrected by using synchronous replication.

Synchronous replication is discussed and analyzed extensively in Chapter 4, <u>Synchronous Replication</u>, of Volume 1 of this series. It involves extending the transaction's scope across the network to include all database copies. Since a transaction is atomic, and since either all changes are made or none are, then at the time of failure, all changes have been made to all database copies or none have. Therefore, no transactions can be lost.

Also, since this networked transaction will lock all copies of a row across the network before updating it (or as part of the update sequence), data collisions cannot occur.

The disadvantage of synchronous replication is that replication is no longer transparent to the application. The application must wait until the transaction has been applied to all copies of the database before it can proceed. This performance impact, called *application latency*, depends upon the technique used for synchronous replication. Three synchronous replication approaches are described below – network transactions, coordinated commits, and replicated lock management.

Network Transactions

The most obvious method for synchronous replication is to simply extend the scope of the transaction across all database copies, as shown in Figure 3-7a. This means that the application or disk subsystem must write to multiple databases, a technique often called *dual writes*. Though this approach can be satisfactory if nodes are collocated, if transaction sizes are small, and if transaction rates are modest, its performance rapidly deteriorates if the nodes are geographically distributed because of the nemesis we discussed earlier – network delays.

Using our previous example of four updates per transaction, we need to send up to eleven round-trip messages over the network – a begin transaction, a read/write for each update, and a two-phase commit. If the nodes are on opposite coasts of the United States, a round trip for a signal requires about 50 msec. Thus, a networked transaction will, in this example, add an additional 550 msec. to the transaction's response time.

Another problem with this approach is that as transaction rates go up, the number of network messages goes up in proportion to the number of nodes in the application network. For large transaction rates, there comes a point at which the network becomes overloaded – a point reached sooner for a WAN than a LAN. The solution is to block messages – a technique that we call *coordinated commits* which we discuss below.

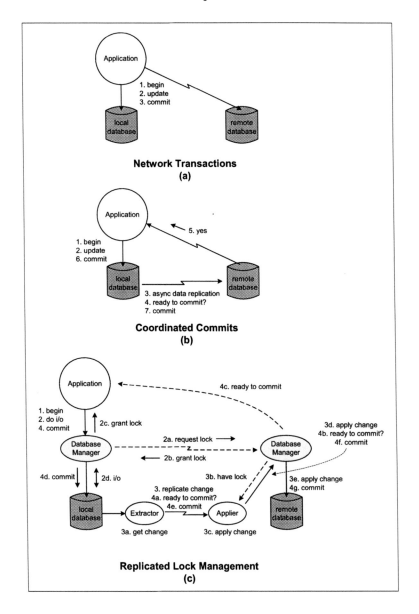

Synchronous Replication Techniques
Figure 3-7

Of course, if the nodes are collocated, network time approaches zero; and network transactions will have less effect except for the additional disk activity at the remote system. However, the network transaction approach suffers from other problems:

- It is an invasive approach. The application must be recoded to include the remote database copies in each transaction. (At the very least, intercept libraries must be created and included to perform this function; or the disk subsystem must support remote writes.)

- It is not scalable. As nodes are added, each additional node must be included in the scope of the transaction. What results is further network traffic and a proportionate increase in application latency.

- As the transaction rate or the number of nodes increases, the network message rate increases in proportion, creating additional communication network load.

- Communication performance may suffer as changes cannot be buffered (as opposed to coordinated commits, discussed next). A change must be sent to the remote site(s) as soon as it is available to avoid impacting the application even further.

- Following a failure, some transactions may be left in an unknown state. If the network should fail just after the target has responded affirmatively to the two-phase "ready-to-commit" query, it can not be determined whether the target received the commit directive or not. Therefore, it is not known whether or not the transaction committed on the target.

- Provision must be made for a node failure. Since all nodes are involved in the transaction, should a node fail, the transaction cannot complete. Thus, a failed node must be excluded from the transaction and re-included when it returns to service.

Coordinated Commits

Another technique is that of *coordinated commits*. The coordinated commit strategy is shown in Figure 3-7b.[48]

With coordinated commits, updates are buffered and propagated via asynchronous replication up until commit time. Therefore, unlike network transactions, the application is unaware of the replication activity. However, rather than immediately sending a commit through the replication channel, as would be done for asynchronous replication, the source system instead sends a *ready-to-commit* query to the target. If the target is ready to commit (typically meaning that it has obtained locks on all rows to be updated), it will respond affirmatively. The source system can then commit its transaction. This commit passes through the replication channel to maintain the natural flow of transactions, and the transaction is applied to the target database.

A network failure during the transaction will cause both systems to abort the transaction, thus maintaining database consistency. Since the target cannot update a row until it has acquired a lock on that row, then data collisions are prevented.

This approach has significantly different performance characteristics than synchronous replication via network transactions. With coordinated commits, the application is not delayed until commit time. At that point, the application suffers a replication latency delay while the ready-to-commit query propagates to the target, followed by a one-way network delay to return the acknowledgement.

---

[48] Holenstein, B. D., et al.; *Collision Avoidance in Bidirectional Database Replication*, United States Patent 6,662,196; December 9, 2003.

Thus, the application is slowed down by a replication latency time plus a one-way network delay. Assuming a replication latency time of 100 msec. and a one-way network time of 25 msec., coordinated commits will add 125 msec. to application response time as opposed to 550 msec. for network transactions in our previous geographically distributed example.

However, if the nodes were collocated, coordinated commits will add a replication latency time of 100 msec. to response time while network transactions will have less impact.

The coordinated-commit approach solves other issues faced by network transactions:

- It is noninvasive since it uses an asynchronous data replication engine for replication. No code changes are required.

- It is much more scalable. As nodes are added, the only additional impact on application performance is the additional commit activity. The ready-to-commit queries and commit directives can be sent to all nodes simultaneously, thus minimizing this effect.

- It provides better communication efficiency. If the system must support 1,000 updates per second, for instance, network transactions as well as replicated lock management would require that 1,000 messages be sent over the network to each node. The coordinated-commit approach allows updates to be buffered, resulting in a significant decrease in the number of messages which must be sent.

- Because it has a restart point marked in the Change Queue, it will automatically repair transactions that ended up in an unknown state following a failure.

- Should a node fail, it must be removed from the commit process. However, this can be done within the synchronous replication engine and thus require no special application modifications.

Replicated Lock Management

Replicated lock management is not distributed lock management. A distributed lock manager is a separate process or node that arbitrates requests for locks and grants locks according to certain policies. With replicated lock management, the requester makes a direct request to a node for a lock on a data item and competes with other requesters for the lock on that node on a first-come, first-served basis. It does not depend upon an intermediate distributed lock manager.

Replicated lock management[49] is a cross between network transactions and coordinated commits. Individual messages are sent to all remote nodes to acquire locks, but the actual database changes are replicated via asynchronous replication. The commit is then coordinated across all nodes.

As shown in Figure 3-7c, when the application requests a lock on a data item (whether it be a read lock or an update lock), the source database manager requests that the target database manager acquire a lock on the same data item. If the target database manager is successful in obtaining the lock, it notifies the source database

---

[49] This discussion of replicated lock management is somewhat academic. Though distributed lock management is a current topic in the cluster community, the authors are not aware of a commercial product that provides synchronous replication using this technique at the time of this writing. However, see Holenstein, P. J., Holenstein, B. D., Strickler, G. E., *Synchronization of Plural Databases in a Database Replication System*, United States Patent 7,003,531; February 21, 2006.

manager that the lock has been granted. If the lock on the target data item is not available, the target database manager will wait to see if it becomes available. After a timeout period, it will notify the source database manager that the lock could not be acquired and the transaction is aborted. Only if both database managers are successful in acquiring the lock is the application granted the lock.

If the lock is granted, the I/O request made by the application is applied to the source database. If the I/O is an update or a write operation, the Extractor of an asynchronous replication engine fetches this change and replicates it to the target database. At the target, the replication engine's Applier verifies that the database manager is in fact holding the lock on this data item. If so, the Applier will send the update to the database manager to update the target database.

After all updates are made, the application will commit the transaction. The source database manager will ask the target database manager if it is ready to commit the transaction. This request is sent via the replication channel, or otherwise marked to indicate the point in the replication stream where it should be inserted, to ensure that the target has received all updates at the time of the request. If the target database manager is ready to commit, it will so inform the source system. At this time, the source system will send a commit to the target system via the replication channel.

One may ask why the commit has to be coordinated across the network if it is known that the target system is already holding locks on all of the data items to be updated. The answer is that just because all of the locks are being held does not mean that the transaction can be committed. For instance, a unique alternate key at the target database might collide and prevent the transaction from committing. The problem is aggravated if the structure of the source and target databases are different. A different key structure or a target database trigger might prevent the target from committing the transaction.

Should the target node or the network fail, the source database manager has to exclude the remote node from further transactions until the node or network is returned to service. At that point, missing transactions must be reconstructed at the target node.

The performance of replicated lock management is poorer than that for either network transactions or coordinated commits. Like network transactions, every I/O operation that requires a lock suffers network latency as the lock request is sent to the target and the lock grant is returned. As with coordinated commits, the application's commit is delayed by the replication latency. Therefore, application latency for replicated lock management is the sum of both the network delays incurred by network transactions and the replication latency incurred by coordinated commits. Its one advantage is that database writes and updates can be blocked and sent via asynchronous replication with its advantages of communication buffering. However, the bottom line is that application latency is greater than that for network transactions or coordinated commits.

Replicated lock management has the following characteristics:

- It is noninvasive to the application, but the database manager must support distributed locks.

- Though data items are buffered for transmission, the requirement to send individual lock requests to all nodes limits its scalability. As the number of nodes is increased, the application latency increases proportionately.

- As the transaction rate or the number of nodes increases, the lock network message rate increases proportional to the number of nodes, increasing communication network loading.

- The replication engine may have a restart point marked in the Change Queue so that transactions that ended up in an unknown state following a failure can be repaired.

Comparison

As has been shown in the descriptions above, the differences between these synchronous replication approaches include the following:

- Depending upon the replication latency, network transactions may be more efficient for collocated systems with transactions of modest size and low transaction rates.

- Coordinated commits are more efficient if the nodes are geographically distributed, transactions are large (such as a batch update), or the transaction rate is high.

- Coordinated commits are scalable. Their application performance impact is independent of the number of nodes in the network and of the size of the transactions. The performance impact of network transactions is proportional to each of these. The performance impact of replicated lock management is proportional to the number of nodes.

- Coordinated commits are noninvasive. This technique can be implemented with no change to the applications. Network transactions generally require that the applications be changed to support the expanded scope of transactions across the network. Replicated lock management requires support to be built into the database managers.

- The performance of replicated lock management is worse than either network transactions or coordinated commits since it incurs the delays of each – the network latency of network

126

transactions and the replication latency of coordinated commits.

An additional issue with synchronous data replication is that it assumes that all nodes are operational. If a node cannot be reached, no transaction can be completed; and the entire system fails. Therefore, means must be provided to detect a failed or isolated node and exclude it from further transactions until it has been restored to service. Database changes made during this interval must be queued and sent to the failed node upon its recovery to resynchronize that node's database. This can be done automatically in a replication engine used for coordinated commits or for replicated lock management, but has to be specially programmed for network transactions.

Deadlocks

All synchronous replication techniques are susceptible to deadlocks in which two nodes are both trying to acquire locks on the same row and stall as a consequence. Two cases of deadlocks exist:

- The first case is the classic case in which the application causes the deadlock. Two applications attempt to lock the same rows in opposite order. Application 1 locks row A and tries to lock row B. Application 2 locks row B and attempts to lock row A. Neither application can obtain its second lock, and a deadlock occurs.[50]

  The deadlock is resolved by having one application release its locks (or both do so and retry at random times). This type of deadlock can be avoided by having all applications use an

---

[50] Replication engines may or may not replicate read locks, but they should replicate update locks in order to keep the databases synchronized.

identical locking protocol which guarantees that locks are acquired in the same order.

Similar deadlocks can occur between an application and a data replication engine's Applier and between Appliers in a multithreaded replication engine. The method to resolve these is as mentioned above.

- The other case is unique to distributed systems and is caused by *lock latency*. Lock latency is the time for a lock at one node to propagate to another node (it is, in fact, the replication latency). Even if both nodes are following the same locking protocol, it is possible for each node to obtain a lock on its local row if these attempts are within the lock latency time. Each will then be unable to acquire the lock on the other node.

  For example, application 1 on node 1 gets a lock on its copy of row 1 at about the same time as application 2 on node 2 gets a lock on its copy of row 1. Neither can acquire the row 1 lock on the other node, and a deadlock occurs.

As with the first case of classic deadlocks, these deadlocks can be resolved by having one or both applications release their locks and retry later at random times. The deadlocks also can be avoided by declaring one node the master node. Locks must be obtained on the master node before they can be obtained on other nodes.

Failure Mechanisms

There are several ways in which a failure can manifest itself in an active/active system. These failures include node failures, database failures, and network failures.

Node Failure

If a node fails, then users at that node no longer have application service. They must be switched to a surviving node so that they can continue.

There are several ways in which users can be switched to a surviving node in seconds, as described in a later chapter. Depending upon the method used, sessions may be lost and have to be reestablished. Certainly, transactions in progress are usually aborted and must be resubmitted.

Besides a few seconds of switchover time and perhaps a need to log on again, the only other impact on a user is a somewhat longer response time since transaction requests now have to be sent to a remote node (though database accesses may still be local to that remote node).

When the failed node is repaired, its database must be resynchronized before users can be switched back to it. Resynchronization is done by the surviving replication engines. While the node was down, the other nodes were queuing changes that must be sent to the downed node. When that node is returned to service, these queues drain their changes to the recovered node. When the queues have been drained to a normal size, the failed node can be considered to be recovered; and its users can be switched back to it.

Database Failure

If a database fails, then a choice must be made. Either the node's database connection must be switched to a surviving database in the network, or the users must be switched to another node.

This is fundamentally a performance decision. The better solution from a performance viewpoint is to switch the users to another node. In this case, transaction requests are delayed by a network delay; but all database activity is local to the user's new node (assuming that this node has a local database).

However, if another node cannot handle the extra load imposed by an additional set of users, then the current node must remain in service and the database connection switched to another (remote) database. In this case, the performance impact is greater since all database activity must go across the network rather than just the transaction request.

In any case, the database must be recovered. Database recovery is discussed later in Chapter 6, <u>Distributed Databases</u>.

Network Failure

If the network is a redundant network, then a network failure will have no impact (except, perhaps, for some performance degradation if the dual networks were sharing the load). All traffic will flow over the surviving network.

However, if a nonredundant network should fail (or in the highly unlikely event of a dual network failure), a node may be isolated from the rest of the application. Again, there are two possible courses of action.

One is to declare the isolated node to be failed and handle the failure as described above.

The other is to let the isolated node continue in service. This is the so called "split-brain syndrome." It will continue to service its users independently of the rest of the application network. The choice of which procedure to use is application dependent.

When the network is restored, all node pairs which have lost a connection will have queued for those connections changes which must now be replicated. Once this backlog of queues at each node has been drained, the system can be considered restored.

However, it is quite likely that during the network outage, many data collisions have occurred if the system is being run in split-brain mode. These must be resolved either automatically or manually before database consistency is restored across the network.

Controlling Database Costs

So far, we have assumed that each node in an active/active system is fully configured. A node is simply a lower capacity version of what the monolithic system would be. If the monolithic system was fault-tolerant, so are the nodes. If the monolithic system had mirrored disks, so do the nodes.

But large disk farms can be very expensive. In fact, they can represent 50% or more of a system's cost. Do we really need to provide a mirrored-disk farm at every node? Having to do so would certainly make active/active systems less attractive because of the cost, though this can be mitigated somewhat by using RAID arrays.

There are, in fact, many options for disk configuration that create an entire spectrum of compromises between reliability, cost, and performance.

Full Mirrored Locality

One could provide mirrored disks or RAID arrays at every node, as shown in Figure 3-8a. In this configuration, there would hardly ever be a database failure at any node. Furthermore, if the node systems were themselves fault-tolerant, then a node failure would be highly unlikely.

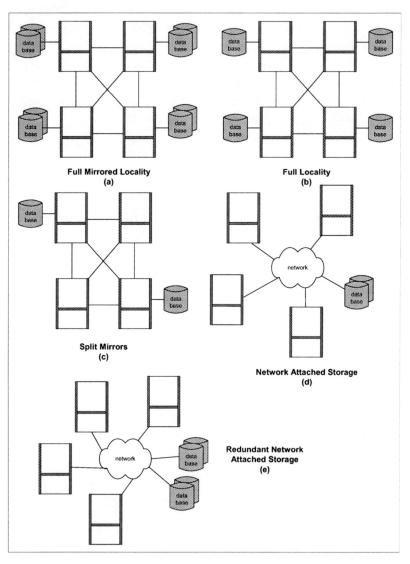

Network Database Configurations
Figure 3-8

Also, every node would have local access to the database, so that performance would be optimized.

The problem with this configuration is, of course, cost. The multiple instances of mirrored databases lead to an expensive system.

Of all of the configurations which we will discuss, full mirrored locality has the highest availability, the highest performance, and the highest cost.

Full Locality

Do we really need a mirrored database at each node? After all, should we lose a database, the node has access to other databases in the network. Thus, we can consider having unmirrored databases at each node, as shown in Figure 3-8b. Connecting to a remote database because of a local database failure will cause reduced performance at the affected node since all database activity must be sent across the network.

Alternatively, if such performance degradation is unacceptable, the users at the affected node can be routed to another node. Now only transaction requests need transit the network. All database accesses are local.

Full locality still has the extreme availability that we are seeking because of the multiplicity of database copies available. In the absence of failure, it also has the best performance. However, it will suffer a greater rate of performance degradation than the fully mirrored system in the event of a database failure. Nevertheless, it will be a lot less expensive than providing a mirrored database at each node.

This configuration maintains high performance and extraordinary availability at a much reduced cost.

Split Mirrors

Do we really need to have multiple copies of the database? After all, mirrored disks are acceptable for fault-tolerant-systems. Disks are much more reliable than processors, and simple mirroring can make a disk subsystem more reliable than the node to which it is connected (except for very large disk farms where there are a lot of failure modes).

Therefore, we can consider having one split mirror[51] in the application network, with a database copy on each of two nodes, as shown in Figure 3-8c. Should one database fail, all nodes (and therefore all users) have access to the surviving database.

This configuration compromises performance since many users connect to a remote database. However, its advantage is a much lower cost while still providing reasonably high availability.

Network Attached Storage

The database copies do not need to be connected to any particular node. Rather, they can be attached to the network so that every node has access to the database via the network, as shown in Figure 3-8d.

Network attached storage is usually redundant, with redundancy provided by RAID arrays. Therefore, this configuration has roughly the same reliability, cost, and performance profile as the split-mirror configuration described above.

---

[51] Not to be confused with the technique in which a mirror is geographically separate from the primary disk in a non-active/active system and is available to a backup system should the primary fail.

134

However, it suffers from one serious fault – it is not disaster-tolerant. Should a disaster of some sort take out the database site, then the system is lost.

Redundant Network Attached Storage

In the network attached storage architecture just described, the lack of disaster tolerance can be solved by providing a second storage system attached to the network, as shown in Figure 3-8e.

This configuration has roughly the same performance characteristics as split mirrors and single network attached storage. However, it has a greater availability and, more importantly, is disaster-tolerant, providing that the two database systems are not collocated. The penalty one pays is higher cost.

The Availability/Performance/Cost Compromise

We have evaluated a variety of active/active architectures and have shown that each has its own availability/performance/cost profile. Each is a compromise between these three factors. Improving one invariably degrades one or both of the others. Figure 3-9 illustrates this.

We conclude this discussion with a review of two of our rules from Chapter 7, The Ultimate Architecture, of Volume 1 of this series:

Rule 29: *You can have high availability, fast performance, and low cost. Pick any two.*

Rule 30: *A system that is down has zero performance, and its cost may be incalculable.*

What's Next?

At the beginning of this chapter, we listed a series of issues faced by active/active systems. We have covered many of them in our review of Volume 1.

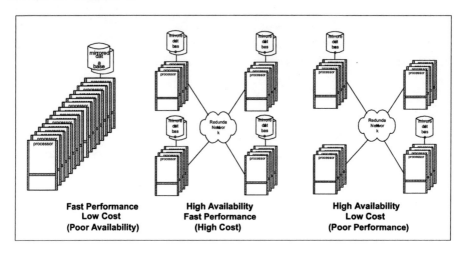

Fast Performance	High Availability	High Availability
Low Cost	Fast Performance	Low Cost
(Poor Availability)	(High Cost)	(Poor Performance)

Availability, Performance, and Cost
Figure 3-9

Several important issues were not explicitly covered in Volume 1, and the next chapters explore these issues and their solutions. Topics include:

- redundant networks
- distributed database recovery
- node recovery
- hardware and software upgrades
- system expansion
- total cost of ownership
- quantitative performance considerations

In the next chapter, we start by looking at a variety of active/active topologies.

Part 2 – Building and Managing Active/Active Systems

In Part 2, we delve into detail concerning the implementation of active/active systems. We first look at various active/active architectures and topologies in Chapter 4.

Underlying an active/active system is a network interconnecting the processing nodes, databases, and users. It is essential that the reliability of the network be commensurate with the desired reliability of the system. Chapter 5 discusses techniques for building highly reliable redundant networks and for recovery from network outages.

Chapter 6 discusses issues with replicated databases and how to recover from database failures. Recovery from node failures is explored in Chapter 7. Chapter 8 describes ways in which active/active systems can be used to eliminate planned downtime.

An important issue in choosing an active/active system is its cost. The total cost of ownership for such a system is the topic of Chapter 9.

Chapter 4 - Active/Active Topologies

"Sites need to be able to interact in one single,
universal space."
- Tim Berners-Lee

Throughout Volume 1 and in the first three chapters of this book, we have focused considerable attention on the theoretical aspects of active/active systems, on their capabilities, and on issues associated with their deployment. Throughout the remainder of this book, we will focus more on the practical aspects of active/active systems in order to provide the reader with a thorough understanding of how to implement them.

We start below with a description of some of the approaches to active/active systems and then conclude with a look at various network topologies supported by these systems.

Architectural Topologies

There are many ways currently being used by enterprises to build what they consider to be active/active systems. Some have many of the desirable characteristics but are not truly active/active. Others provide true active/active capabilities but are subject to certain limiting conditions in order to avoid some of the critical issues that can occur with active/active architectures. Still others are fully active/active, and these must deal with the variety of issues characteristic of such systems. We look at many of these architectures and comment upon them.

141

Database Partitioning

The following architectures are designed to avoid data collisions when using asynchronous replication. They are based on partitioning the database so that updates to a row are always made to a specific database copy before replication, thus avoiding data collisions.

[Note: This chapter has been formatted so that each major section starts on a fresh page in order to better highlight the major topics.]

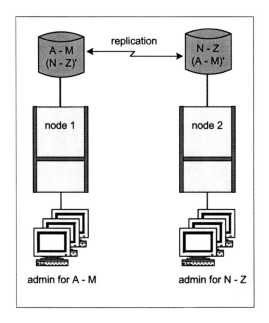

Database Partitioning by User Locality
Figure 4-1

User Locality

Perhaps the simplest method of database partitioning is based on user locality. This technique is applicable if there is a natural partitioning of data by user activity.

For instance, consider a two-node active/active system as shown in Figure 4-1. Assume that one group of users only updates customers A to M, and another group updates only customers N to Z. Then the database can be partitioned between the two nodes such that the primary partition for customers A to M is on Node 1 and the primary partition for customers N to Z is on node 2. Likewise, all users associated with customers A to M are connected to Node 1, and all users associated with customers N to Z are connected to Node 2.

Each node backs up the partition on the other node by using asynchronous data replication. As a consequence, any updates made to a particular row are always made to the same database copy and replicated to the other copy. Therefore, no collisions can occur.

This model is easily extended to multiple nodes in an application network by having each node back up only one other partition. For instance, if there are four nodes, the database is partitioned into four partitions. Each node carries one-fourth of the database as its primary partition and backs up the primary partition of one other node. Thus each node carries half the database, and there are a total of two database copies in the application network, as one would want in an active/active configuration.

In this configuration, though any particular user will update data on only its home node, it has access to all data across the network. In a two-node system, all data access is provided by the user's home node since the entire database copy resides there. In a multi-node system as described above, local data access is provided for two partitions, but data in other partitions must be obtained by accessing a remote node. If local data access is desired for all data, then all partitions could be resident on all nodes. An update to a primary partition would then be replicated to all nodes. The downside of this approach is cost. If there are *n* nodes, then there are *n* copies of the database (a cost/performance tradeoff).

There are many cases in which this configuration *almost* works. In these cases, the bulk of a user's updates are to his home node. However, occasionally he may have to update data in a foreign partition. In this case, there are several solutions:

- Update the local copy and deal with the (hopefully, occasional) data collisions as discussed later under <u>Asynchronous Replication – Data Collision Detection and Resolution</u>.

- Reach out across the network and make the update to the data's primary partition in a remote node either directly as a part of the transaction or as a subtransaction generated as part of the parent transaction.

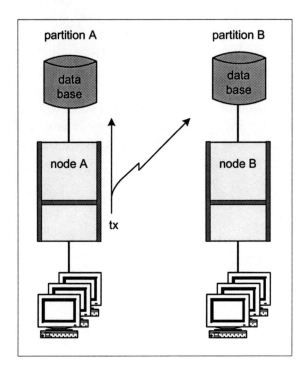

Database Partitioning by Remote Transactions
Figure 4-2

Remote Transactions

With this technique, the database is physically partitioned between the various nodes in the application network, as shown in Figure 4-2. All of the nodal databases must be taken in total to comprise the entire application database. As a result, any particular row in the database is resident at only one node.

Any transaction can be entered by a user at any node. Updates to rows resident at that node are made locally. Updates to rows on other nodes are made over the network. All updates associated with a transaction are included within the scope of the network transaction so that either all updates are made across the network, or none are.

This is an active/active configuration in the sense that all nodes are active and that any transaction can be processed by any node. In addition, there can be no collisions since each data item resides in only one place in the application network. Furthermore, this architecture lends itself to scaling, as the database can be further partitioned by adding additional processing nodes.

However, it misses the active/active criteria since there is only one copy of the database in the network. Should a node or disk subsystem fail, the entire system is down since a portion of the database is no longer available. This configuration is certainly not disaster-tolerant.

Unless the application has a great deal of locality, or unless the databases are closely located with very high-speed access between them, communication latency could cause serious response-time problems. If transactions primarily update data local to their own nodes, communication channel propagation time may not be a problem. However, if a transaction can typically span multiple nodes, and if the nodes are far apart, then channel propagation time may be a major factor in the response time.

One complexity occurs if a single transaction can update data in multiple partitions. In this case, the transaction will have to be broken up into a series of subtransactions, each updating a particular database partition. Typically, the application cannot do this as it cannot guarantee the atomicity of the transaction (either all subtransactions commit or none do). The management of subtransactions must generally be a capability of the database manager or of the transaction manager.

The above configuration is similar in many respects to a federated database, in which a single system view is provided across a set of

147

distributed databases. Federated databases are discussed in more detail in Chapter 16, <u>Related Technologies and Drivers</u>, in Volume 3.

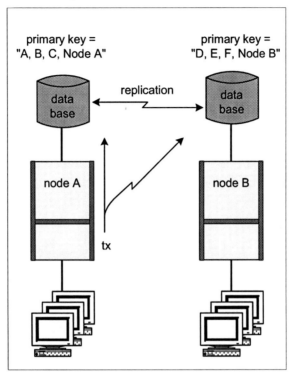

primary key =
"A, B, C, Node A"

primary key =
"D, E, F, Node B"

replication

data base

data base

node A

node B

tx

**Database Partitioning with Database Modification
Figure 4-3**

Database Modification

A replicated database can be partitioned by qualifying each row with a node identifier, as shown in Figure 4-3. For instance, if the primary key for a row would normally be [A, B, C], then when the row is first created, it would be given a primary key of [A, B, C, Node ID], where Node ID might be the identifier of the node on which the row was created. Alternatively, the node identifier could be placed in its own column in each row. Let us call this the *source node identifier*.

This same concept can be applied to other scenarios, such as which node *owns* that data item. Furthermore, the identifier could change over time as the partitioning strategy changes.

Since the database is replicated, the database copies are typically synchronized via data replication or network transactions. Thus, each row in each database copy carries with it its source node identifier.

When a node needs to update a row, unless that row resides locally, the node will reach across the network and update the row on the row's source node. This change will then be replicated to all nodes, including the node initiating the change.

Since a row can only be updated in one particular database copy (i.e., the copy on its source node), there will be no data collisions.

This is truly an active/active system. Transactions can be processed by any node, and the application's database is replicated. Furthermore, this configuration is data-collision free.

However, the system is subject to performance degradation due to communication latency unless transactions tend to be local (that is, most updates are to rows for which the processing node is the source node).

Perhaps of more concern is that for existing applications, the database must be restructured to provide source node identification for each row. Additionally, the application itself may have to be restructured to supply the node identifier when accessing the database.

This architecture may be suitable, for instance, for certain telco environments, such as logging call detail records in a cell phone application.

An interesting extension of this concept applies to the generation of unique numbers required by the application. They might include customer numbers, invoice numbers, sequence numbers, random numbers, and so forth. If each node were to simply generate its own numbers, there would surely be duplication across the network, such as multiple customers with the same customer identifier. One solution to this problem is to simply append (or pre-append) the identifier of the generating node to the generated number. Then the number is guaranteed to be unique across the network.

An alternative solution is to have each node generate id numbers that start with its node number and that then increment that number by the number of nodes in the system. For example, for a three-node network, Node 1 generates id's 1, 4, 7, ..., Node 2 generates id's 2, 5, 8, ..., and Node 3 generates id's 3, 6, 9, ...

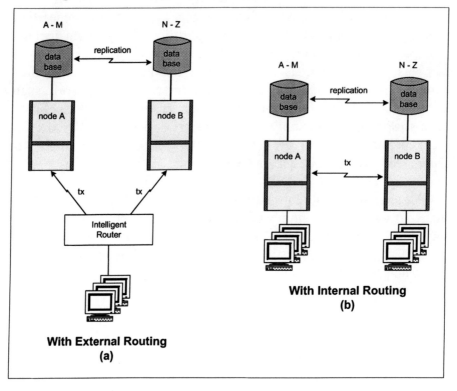

Database Partitioning by Data Content Routing
Figure 4-4

Data Content Routing

A powerful way to partition the database is at the data content level. For instance, customers could be broken up by name, account number, or customer identifier, with specific customer blocks residing on specific nodes as their *source node,* as shown in Figure 4-4. Now the source node for a transaction can be determined by the content of the transaction, and the transaction can be routed to that node. This is known as data content routing. Updates to a source node are replicated to the other database copies.

Routing transactions can be done in several ways. An external intelligent router can be used to receive transactions and to forward them to the appropriate node (Figure 4-4a).[52] Alternatively, each transaction can be preassigned to a node or sent to the node most local to the submitting user; and the node can reroute it if necessary (Figure 4-4bb).

These configurations are especially powerful if there is a great deal of locality in the application; that is, users typically submit transactions that will be processed locally by their nodes.

Depending upon the application type and data-partitioning method used, partitioning by data content routing can potentially break down for complex transactions that require access to nonlocal database items. For instance, if a customer transaction includes part numbers, those part numbers can often have a source node that is remote. To avoid collisions, either the application will have to update those rows over the network, or it will have to generate subtransactions to be sent to other source nodes. In either case, the application will have to be modified; and response time will suffer.

---

[52] In telco applications, this is sometimes called a "mated pair."

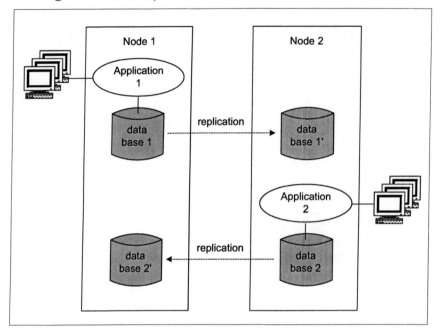

Database Partitioning by Application
Figure 4-5

Application Routing

Data can be partitioned by application, as shown in Figure 4-5. With this technique, different applications are running in each of the different nodes and are using their own local databases. Therefore, no collisions will occur. Each application database is backed up by another node in the application network. Should a node fail, that application can then be brought up on the node that was hosting its backup database.

Application routing configurations are not really active/active systems since the nodes are not symmetric. They are not running the same applications, and they are not accessing a common database. However, the nodes are all being effectively used; and recovery from a failure is straight-forward though not measured in seconds as with an active/active system. Rather, this configuration is an extension of the active/backup architecture, in which an application database is backed up by a copy in a remote system. Following a failure, the backup database must be brought into consistency; and the application must be started on the backup system.

Software Replication

Most of the techniques described above share the active/active characteristics of high availability and fast recovery. However, not all provide the other advantages that can accrue from a true active/active architecture.

These other advantages are all realized by building active/active systems around data replication, which has been the focus of this series.

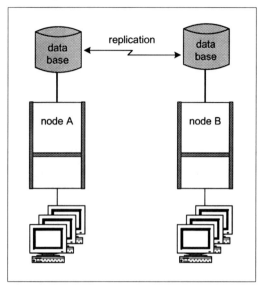

Asynchronous Replication
Figure 4-6

Asynchronous Replication

Asynchronous replication (Figure 4-6) replicates changes from a source database to a target database with little or no impact whatsoever on the source application. As we have discussed, there are many advantages of using asynchronous replication to build active/active systems relative to other approaches:

- There is no performance penalty. Replication is done asynchronously with respect to the application.

- It is generally nonintrusive. Neither applications nor the database have to be modified.

- System integration is done at the database level, not at the application level. In addition to not having to modify the

157

applications, system security can be more easily managed; and multiple database connections do not have to be supported by the applications.

- It is scalable. It is relatively easy to add nodes to gain additional capacity.

- All purchased capacity is available for use by the applications. There need be no idle backup nodes in the system.

- Planned downtime can be eliminated.

- Disaster tolerance comes "for free" if the nodes and database copies are geographically distributed.

However, asynchronous replication brings with it several issues, as shown in Figure 4-7, all of which have been described previously in Chapter 3 of this volume:

- Data oscillation must be avoided. It must be ensured that a replicated change is not returned to the source system (ping-ponging) (Figure 4-7a).

- Any data in the replication pipeline may be lost following a failover (Figure 4-7b).

- Data collisions may occur should the same row be changed independently at nearly the same time at two different nodes, thus corrupting the database (Figure 4-7c).

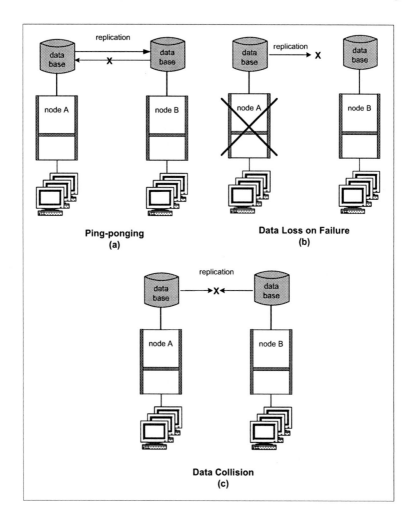

Asynchronous Replication Issues
Figure 4-7

There are existing solutions for ping-ponging,[53] and any data replication engine capable of bidirectional replication must include this protection.

The data loss and data collision problems are solved if synchronous replication, described later, is used. However, synchronous replication imposes a response time penalty (which we have called *application latency*[54]) on the application.

When asynchronous replication is used, data loss and data collisions can be minimized by reducing the replication latency of the replication engine. In addition, several architectures that avoid or minimize data collisions have been described above. There are more general steps that can be taken to identify and resolve data collisions when using an asynchronous replication engine, as discussed next.

Single Entity Instance

Let us call the subject of a transaction an *entity*. An entity might be, for instance, a credit card, a bank account, a hospital patient, or a telephone number.

If there can only be a single physical entity instance that is represented by the database information, then there can be no possibility of a collision. For instance, in a hospital patient record system, there is only one instance of each patient; and that patient will only physically be at one hospital at a time. If a single node can handle that hospital's load, there should only be one node submitting transactions at a time for that patient; and data collisions will not

---

[53] Strickler, G., et al.; *Bidirectional Database Replication Scheme for Controlling Ping-Ponging,* United States Patent 6,122,630; Sept. 19, 2000.
[54] See Chapter 3, Building Active/Active Systems.

occur.[55] A single physical entity is one which is "owned" by only one node at a time.

Multiple Entity Instances

If there can be multiple physical entity instances which are represented as a single entity in the database, then data collisions can certainly occur. This is because competing updates could come from different nodes. An example of multiple entity instances is an ATM card. A husband and wife might both have access to the same account via their ATM cards (the two ATM cards are two physical instances of the same database entity, the account); and they can, in principle, make simultaneous transactions through different nodes. This may lead to a possible collision.

An extension to the above example is the posting of payments. In this case, the posting application is another instance of the ATM card as it also can post transactions to that ATM account. Similar examples are joint credit cards and joint brokerage accounts.

Data Collision Detection and Resolution

Whether data collisions can be avoided or will occur and then need to be detected and resolved is generally a function of the network architecture and the data replication approach used (asynchronous or synchronous replication). When using synchronous replication, data collisions will be avoided. We also have discussed certain partitioning architectures that avoid data collisions when using asynchronous replication. However, in general, when using asynchronous replication, data collisions may indeed occur. Handling these is discussed next.

---

[55] Typically, all patient results, doctor comments, and administrative information would be entered into the node used by the hospital to which the patient is currently admitted. Of course, if the hospital is running several nodes, or if information is being entered at nodes at other hospitals, this may not be true.

The first step in handling data collisions is to know when they occur. Some replication engines provide *collision detection* by sending some sort of source row version identifier with the change being replicated. If the target row version is different, then a collision has occurred. Version identifiers might be a date/time (if the nodes are time-synchronized), a row version sequence number that is incremented for each change, or the *before* image of the source record itself (the most generic version identifier).[56]

To understand the techniques that can be used for *collision resolution*, it is first important to categorize the several types of data collisions that can occur. This depends upon the database operation, as shown in Table 4-1. Here, the *Source Operation* column represents what the source node did. The *Target Database State* column represents the state of the target database before the source operation is applied to it. The *Target Action* column shows the outcome of the source operation on the target database given the current state of the target database.

For example, if the source node inserted a row, and if that row did not exist in the target database, that row would be inserted. However, if the row already existed in the target database, then a collision would have occurred.

If a collision is detected, it must be resolved. In many cases, this can be automated by embedding business rules into the replication

---

[56] When detecting colliding changes, an interesting problem is the "ABA" problem. Node 1 changes the contents of a row from A to X. At the same time, Node B changes the contents of the same row from A to B and then back to A before the update from Node 1 is received. If before images are used, the Node 1 update is allowed and applied at Node 2, and the database is consistent, but some change activity (the row B update) is not detected (at Node 2). In some applications, this is not acceptable, and another row versioning scheme must be used such as a date/time stamp or a sequence number, as these will detect such "oscillating" row change data values.

engine (for instance, via scripting or user exits) to instruct it how to resolve the collision. If the collision cannot be resolved automatically, then it should be logged for later manual resolution.

Source Operation	Target Database State	Target Action
insert	row does not exist	apply insert
	row exists	collision
update	row exists, and the source row version is the same	apply update
	row exists, and the source row version is different	collision
	row does not exist	collision
delete	row exists, and the source row version is the same	apply delete
	row exists, and the source row version is different	collision
	row does not exist	collision

Collision Types
Table 4-1

A typical set of business rules for collision resolution might be the following:

- If an inserted row already exists, convert the insert to an update and apply it.

- If an updated row exists, but the source row version is different, resolve via business rules.

- If an updated row does not exist, convert the update to an insert and apply it.

- If a deleted row exists, but the source row version is different, delete it (alternatively, ignore the delete).

- If a deleted row does not exist, ignore the delete.

These simple collision rules may lead to database divergence since the same operation performed on the source database may not be performed on the target database. For example, consider an active/active system with two nodes, each with its own database copy. Assume that each node simultaneously (within the replication latency time) receives a valid insert for the same row but with different data in the nonkey columns. In this case, each node will apply the insert to its database and will then replicate the insert operation to the other node. Since each node now has a row with this key, the replicated insert will be converted to an update and will be applied to the now existent row. The result is that the copies of this row in the two databases do not match, and the databases will begin to diverge. Clearly, a higher form of data collision logic is required in order to keep the databases properly synchronized.

Some examples of higher forms of collision resolution logic are described below.

Relative Replication

For numeric fields, collisions may be resolved via *relative replication*. With this technique, rather than replicating the entire updated row, only the operations on numeric fields are replicated (e.g., add 5, subtract 30).

For instance, assume a numeric field in a row had an initial value of 50 on both nodes. Node A applies an update that adds 12 to that row, and Node B applies an update that subtracts 6 from that row. With absolute replication, Node A will replicate a value of 62 to Node B; and Node B will replicate a value of 44 to Node A. The value of this field will be different in the databases of Nodes A and B, and both will be wrong.

With relative replication, Node A updates its field value to 62 (50+12) and replicates the operation "add 12" to Node B. Node B updates its field value to 44 (50-6) and replicates the operation "subtract 6" to Node A. Upon receipt of that operation, Node A will update its field value to 56 (62-6). Likewise, upon receipt of the replicated operation "add 12" from Node A, Node B will update its field value to 56 (44+12). In the end, both databases agree; and both are correct.

Note that addition and subtraction are commutative, as are multiplication and division, in that they can be done in any order. From a relative replication viewpoint, $(10 + 3) - 5 = (10 - 5) + 3 = 8$ and $(10 \times 3) / 5 = (10 / 5) \times 3 = 6$. The order in which the operations are performed does not matter. However, these two sets are not mutually compatible when it comes to relative replication. If the initial value of a field is 10, and one system multiplies this field by 3 while the other adds 15, order is important. $(10 \times 3) + 15$ is not the same as $(10 + 15) \times 3$ (45 and 75, respectively). Therefore, these operations cannot be intermixed when using relative replication.

165

Relative replication is a form of the use of *commutative transactions.*[57] Commutative transactions are those that can be applied in any order, and the database will end up in the same state. Timestamped appends to a table are a form of commutative transactions.

Row Content

If relative replication is not an option (for instance, the update is to a text field), then the change to be accepted can be determined by the *row contents*. For instance, if each row contains a timestamp field, the change with the latest (or earliest) timestamp can be chosen. If each row contains a sequence number, the larger (or lesser) sequence number can be chosen.

As an example, let us assume that rows carry a timestamp field and that the latest row (i.e., that with the most recent timestamp) will be accepted. If an application on Node A applies an update with a timestamp of 09:17:33, it will replicate the row with this timestamp to Node B. If, prior to receiving Node A's replicated update, an application on Node B applies an update to the same row in its database copy with a timestamp of 09:17:34, this update will be replicated to Node A. Both nodes will detect a collision (perhaps since the before images do not match) and will select the row timestamped 09:17:34 as the proper update to keep.

Node Precedence

The change to be accepted can also be determined by *node precedence*. Using this technique, each node is given a precedence order. When a data collision occurs, the change submitted by the node with the highest precedence is accepted.

---

[57] See Gray, et al.; *The Dangers of Replication and a Solution,* ACM SIGMOD Record; 1996.

For example, in a three-node active/active system, Node A might be given precedence 1 (the highest), Node B be given precedence 2, and Node C be given precedence 3. Should the nodes receive colliding changes from Nodes B and C, the change from Node B will be accepted.

Combination

Often, a combination of the above techniques is appropriate. For instance, numeric collisions can be resolved via relative replication, and text collisions can be resolved by accepting the later of the changes based on row content or by using node precedence to select the winning change.

Business Algorithm

In certain cases, the resolution of a collision requires some sort of complex business algorithm that does not map into any of the above techniques. In these cases, specialized collision resolution procedures may be implemented via scripting facilities provided by the data replication engine. Alternatively, specially coded algorithms can be included as user exits and bound into the replication engine.

Using this technique, for instance, a collision can be treated like a deadlock. Either the transaction is aborted and both users are notified to resubmit the transaction, or the involved nodes each wait a random amount of time and resubmit the change.

Designated Master

A variation of Node Precedence is to designate one node as the master node. All nodes make their changes to their own database copy and then replicate those changes to the master node. The master node then replicates each change to the other nodes in the application network, including the node that originated the change.

Should the master node receive a colliding change from two or more of the other nodes, it can determine the winner, perhaps by using a "first-to-update" rule or by using one of the above algorithms to resolve that collision.

Using a master node solves the Database of Record problem since the database on the master node is by definition the Database of Record.

The designated master procedure has some further implications when applied to active/active systems. First of all, all transactions except those originated by the master node will be slowed since updates must travel across the network through the master to reach the other nodes. Secondly, provision must be made to promote another node to master should the master fail.

Master/Slave

The Designated Master configuration can be used in a somewhat different way by requiring that all updates be made directly to the master node. Rather than making the update to the local copy of the database and replicating it to the master node, updates are instead made across the network directly to the master database. The database changes are then replicated from the master node to all of the other nodes, now known as the slave nodes.

Using a Master/Slave configuration, no data collisions can occur. However, applications running in the slave nodes pay a performance penalty since all updates must occur across the network instead of locally.

Self Correcting

Rather than attempting to resolve collisions, one could simply allow the collisions to occur and let the databases diverge. It may not matter that the databases diverge for a while if subsequent application processing may force the databases into convergence.

For example, should the database copies diverge, it may only be a matter of time (perhaps minutes to hours) before they converge again due to other noncolliding updates to the same data. These updates will be applied to all copies of the database, bringing those rows into synchronism across the network.

Alternatively, a periodic refresh could be used to resynchronize the databases.

Manual Correction

For many applications, data collisions may be quite rare[58] and their resolution complex. In this case, manual review and action may be the best approach to collision resolution.

Data Collision Logging

In any event, data collisions and their resolutions could be logged for later review and possible additional action.

---

[58] See Chapter 9, <u>Data Conflict Rates</u>, in Volume 1 of this series, *Breaking the Availability Barrier: Survivable Systems for Enterprise Computing*, for an analysis of data collision rates.

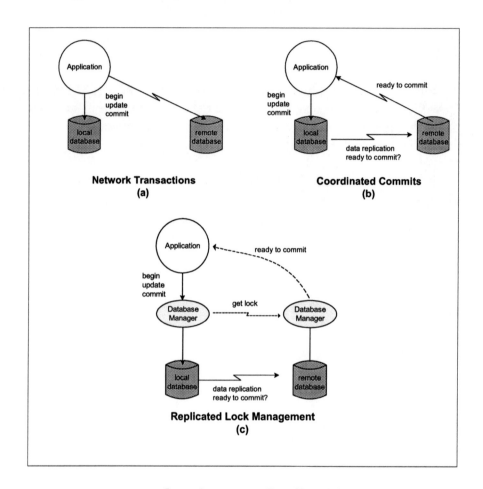

Synchronous Replication
Figure 4-8

Synchronous Replication

Data collisions as well as data loss following a failure can be eliminated by the use of synchronous replication.[59] Using

---

[59] See Chapter 3, <u>Building Active/Active Systems</u>

synchronous replication, each node must first acquire locks on all rows to be updated in all database copies across the application network. Once this has been confirmed, then the transaction can be committed on each database copy.

Since all rows are locked by the transaction either before or as updates are applied, no other application can modify these rows; and no data collisions will occur. Furthermore, either all updates are made to all database copies, or none are. Therefore, no data will be lost as the result of a failure (although it is possible for a transaction to be hung in an unknown state while awaiting later resolution[60]).

We have discussed synchronous replication in some detail in Chapter 3 of this Volume. In that chapter (and in more detail in Volume 1 of this series), we described three possible approaches to synchronous replication, briefly reviewed in Figure 4-8 – network transactions (Figure 4-8a), coordinated commits (Figure 4-8b), and replicated lock management (Figure 4-8c).

With network transactions, a global transaction is started across the network that includes all database copies in its scope.

With coordinated commits, changes to the source database are replicated to the target database via asynchronous data replication, transparent to the application. However, the transaction is not committed until all database copies have indicated that they are ready to commit.

With replicated lock management, a lock on each data item is first acquired across the network from all database copies. Updates are

---

[60] In effect, the asynchronous data-loss window changes into a data-uncertainty window for synchronous replication due to the nature of the coordinated commit or two-phase commit protocols. The transaction data is indeed safe-stored at the target; but depending upon the nature of the failure, the transaction final state (i.e., commit or abort) may not be known.

then replicated via an asynchronous replication engine. Once all updates are made, the transaction will not be committed until all database copies have indicated that they are ready to commit.

There are some issues with synchronous replication. One is application latency. Since the source transaction, and therefore the source application, must wait for all remote commits to be completed (or at least for the remote nodes to indicate that they are ready to commit), the completion of the transaction at the source will be delayed and will lead to an increase in application response time. Whether network transactions or coordinated commits perform better depends upon the aggregate of the communication channel delays, the size of the transaction, the loading on the source system, and the replication latency of the replication engine.

With network transactions, each start transaction, update, and commit must flow independently as a round-trip message over the communication channel. With coordinated commits, the source application's transaction runs at full speed during transaction processing. Then at commit time, the application must wait a replication latency time while the "Ready to Commit" query is sent to the target and the response received. If the replication latency is less than the aggregate communication delay needed by network transactions, then coordinated commits are the preferred method, and vice versa.

Replicated lock management's performance is poorer than either network transactions or coordinated commits since it incurs both the network delays of network transactions and the replication latency of coordinated commits.

Furthermore, coordinated commits will use the communication channel more efficiently since the channel can send changes in large blocks rather than one at a time, as required by network transactions. Also, less processor time will be used because the number of blocks to be processed will be less (communication block processing consumes valuable interrupt processing time). This effect becomes more predominant as transactions rates increase.

In addition, implementing coordinated commits into an existing application is generally noninvasive, whereas network transactions generally will require application source code changes or database changes to include all database copies in the scope of the transaction.

This leads to an enhancement of Rule 17, stated earlier in Chapter 4 of Volume 1:

Rule 17 (enhanced): *For synchronous replication, coordinated commits using data replication become more efficient relative to network transactions and replicated lock management under a transaction manager as transactions become larger, as communication channel propagation time increases, or as the transaction load increases.*

One other synchronous replication issue can occur when a node fails. If a node cannot commit a transaction, then no transaction can be committed. The application has effectively crashed. Therefore, the replication strategy must exclude a failed node from the scope of further transactions. Once the node is repaired, its database must be resynchronized.

With coordinated commits, database resynchronization is simply done by draining the replication queues which have been built during the outage. Network transactions will typically require a custom utility to do this or must otherwise reload the database.

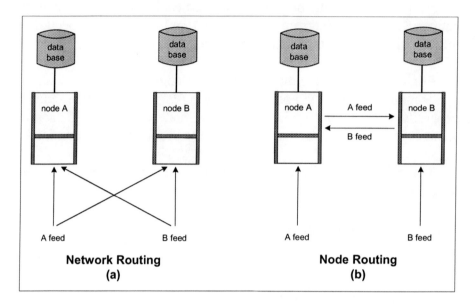

Transaction Replication
Figure 4-9

Transaction Replication

Another topology for active/active-like processing is to feed all transactions to all nodes, as shown in Figure 4-9, and to have each node process all transactions. In this way, each node will update its own copy of the database; and no data replication is needed.

Transactions can be sent to all nodes via the network (Figure 4-9a). Alternatively, a transaction can be sent to the node local to the submitting user; and that node can forward the transaction to the other nodes (such as via a middleware facility such as IBM's MQ Series), as shown in Figure 4-9b.

In any event, the point is that each node processes all transactions.

This approach has some important limitations. First of all, it is not scalable. Each node must process all transactions. The load can only grow as large as that which the smallest node can handle.

Secondly, it is subject to its own form of data collisions. For example, there is no guarantee that a transaction will be processed in the same way at each node. In addition, they may not be processed in the same order since they may be arriving over different paths.

For example, let us consider an inventory application. A particular part has an inventory of ten units. Two transactions against the part are generated nearly simultaneously – transaction A for an order of eight parts and transaction B for an order of four parts.

Node A receives transaction A first. It will reduce inventory by eight. Upon receipt of transaction B, it will reduce inventory to zero and perhaps will backorder two units for Customer B.

Node B, meanwhile, gets transaction B first. It will reduce the quantity to six. Upon receipt of transaction A, it will reduce the quantity to zero and perhaps backorder two units for customer A.

Alternatively, Node A could receive a transaction just prior to the batch cutoff time and be included in today's activity. Node B could receive this transaction just after the cutoff time and thus include it in tomorrow's activity.

Depending upon the application, it may be difficult or impossible to detect such collisions. Often, they can only be detected by verifying that both databases are identical. Transaction replication, however, may be suitable for applications that only insert rows, such as those which log call detail records in a telephony application. In this case, collisions will generally not occur.

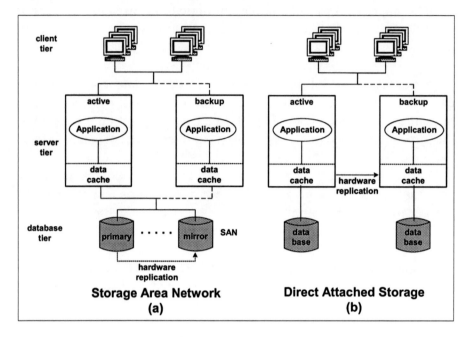

Hardware Replication
Figure 4-10

Hardware Replication

Throughout Volumes 1 and 2 of this series, we talk about *application* transaction-level software replication.[61] Replication is driven by a queue of change events which are sent to the target system and applied to its database. This form of replication is able to replicate at the transaction level in that transaction control information is also replicated, and updates are applied to the target as transactions. That is, either all updates in a transaction are applied, or none are. As a result, the target database can always satisfy all of the requirements of referential integrity and database consistency and consequently can be

---

[61] Application transactions are what matter from a business viewpoint. Hardware replication, as we will see, does not preserve business transaction boundaries.

used by other application instances. This is a requirement for active/active systems.

However, there is another form of replication in common use. It is called hardware replication. Hardware replication is found predominantly in enterprise storage systems such as storage area networks (SANs) and network attached storage (NAS). As we shall see, hardware replication does not maintain a consistent database and is not suitable for active/active systems. However, it can be very useful for replicating between an active system and a backup system. It is important to understand why this is so.

In disk systems, writes to a physical disk are usually complete blocks (sometimes tracks), where a block comprises one or more sectors. Hardware replication replicates blocks (as opposed to software replication, which replicates changes). As a block is written to a primary disk, that block is sent to the backup disk, where it is similarly written. By replicating blocks in this manner, the writing to the target disk is usually very fast.

Consider the tiered representations of an application as shown in Figure 4-10. Figure 4-10a shows an active/backup pair of nodes which use a SAN for their data storage. In this case, hardware replication is done by the disk subsystem. Whenever a disk block is flushed from the primary node and written to a primary disk, that block is replicated and written to the backup disk by the disk controller.

Figure 4-10b shows the same system but using direct attached storage instead. In this case, "hardware" replication is often done at the cache level by a software facility. When a block is flushed to a disk by the primary node, that block is sent to the backup system, which will flush it to its disk.

177

Hardware replication may be either asynchronous or synchronous.[62] With asynchronous hardware replication, as with asynchronous software replication, the application is unaware of the replication activity. However, with synchronous replication, the write at the primary node does not complete until the block write at both nodes has been completed.

The time to complete a synchronous write is very dependent upon the communication channel latency between the active and backup nodes. Therefore, with synchronous hardware replication, there is generally a distance limitation between the two nodes – typically in the order of 100 kilometers. With asynchronous hardware replication, there is generally no distance limit.

Note that all block writes are replicated. This means that the source disk and the target disk are always exactly the same. Therefore, they cannot be heterogeneous. They must be identical down to the format of data on the disk.

The fact that they are identical means that should the source disk suffer data corruption, the corruption is replicated exactly as is to the target disk. This may prevent the target disk from being opened for recovery following a failure.

However, block writes mean that hardware replication will replicate any kind of data – not only transaction data but also transaction logs, configuration files, control files, source and object files, and other unstructured data.

One problem with hardware replication is that data is generally not replicated until it is flushed from the node's cache. For transactional systems, there are two kinds of data of interest – the

---

[62] In some implementations, semisynchronous hardware replication may be used in which a block is replicated asynchronously, but the next block cannot be written until the previous block has been acknowledged.

database itself and the transaction log. The transaction log contains before and after images for all database operations as well as transaction control data indicating which changes were part of which transactions, which transactions have been completed, and which are still in progress.

The transaction log is typically flushed after the completion of each transaction and therefore will be fairly current at the target. However, data blocks are generally flushed much more infrequently; and different disks in the system's disk farm are flushed independently. The result is a great deal of inconsistency at the target. Index blocks might be missing, block splits may be incomplete, child rows may have no parents, and so on. The physical database does not equate to the logical database, and any semblance of referential integrity is totally sacrificed. It is because of this inconsistency and other problems such as the circular replication of changes that the target database cannot be actively used and why hardware replication is not applicable to active/active systems.

Furthermore, following a failure, it is not known how old the data in the database is. Though the transaction log may indicate the completion of many recent transactions, there is no guarantee that the associated data has been flushed and replicated. Much of this data was still in the cache of the failed system at the time of failure and has been lost. These transactions must be reconstructed from the transaction log. Unfortunately, there is no way to tell how far back in the transaction log one has to go to reconstruct transactions whose data has not yet been replicated (in fact, has a necessary part of the transaction log been rolled off?). The result is a very complex recovery process, and this process may require a potentially long recovery time. This is in sharp contrast with software replication, which guarantees that all changes will be applied to the target database when a commit is replicated.

The complex recovery procedures required by hardware replication can by somewhat mitigated by turning off cache and using cache write-thru to write data changes to disk immediately, and coupling this with the synchronous replication of block changes. Using this technique, all data is written to the target system in the same order as it was changed on the source system. This technique, of course, imposes significant performance degradation since cache is not used to buffer writes; and each block update must wait for the block to be successfully replicated to the target system. Furthermore, there is still a recovery issue since replication is not by transaction. Determining which changes are applicable to completed transactions and which are not is a complex problem.

Hardware replication imposes significantly more load on the communication channel. Rather than simply replicating rows (typically hundreds of bytes), blocks (typically thousands of bytes) must be sent to the target system. However, the chance that a replicated data block may contain several changes mitigates this effect to some degree.

In summary, following are some of the advantages of hardware replication:

- There is no CPU overhead if replication is done at the controller level.
- Target updates are fast since blocks are overlaid rather than updated.
- If asynchronous replication is used, there is no impact on the application.
- If synchronous data replication is used, there is no lost data following a failure of the active system (zero RPO).
- Nonstructured data is replicated.
- It is relatively easy to configure.
- It is suitable for disaster recovery configurations using an active system backed by a backup system.

Limitations of hardware replication include the following:

- There is no concept of database consistency nor of referential integrity. Hardware replication, therefore, cannot be used for active/active systems.
- The primary and backup systems must generally use identical database hardware and software.
- Recovery from a failure is a complex and lengthy task and leads to large recovery times.
- A significant amount of communication bandwidth is required.
- The backup disk is not usable even for reporting as it is highly inconsistent due to missing data not yet replicated.
- Data corruption of the source database is replicated to the target, perhaps preventing it from even being opened for recovery.
- Synchronous replication limits the distance between nodes (typically about 100 km).
- With synchronous replication, changes to the primary disk volume are slowed to the speed of the communication link and backup volume.

Recall our Rule 49 from Chapter 3:

Rule 49: *Replication engines that violate referential integrity are rarely, if ever, suitable for active/active implementations.*

We can now add a corollary to that rule:

Rule 50: (Corollary to Rule 49) - *Hardware replication is not suitable for active/active applications because it does not provide referential integrity.*

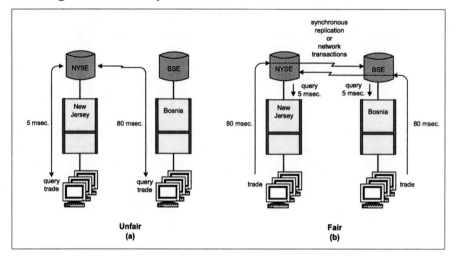

Fairness
Figure 4-11

The Fairness Doctrine

When data is distributed across the network, the amount of time it may take to access or update that data is usually a function of the distance between the requesting application and the database containing that data, as well as the communication technologies in use. This is nothing more than a response time aggravation in most instances. However, there are certain critical applications in which differences in data access time can be crucial.

An excellent example of this problem is an auction market. In such markets, an item of some sort is put up for sale; and bidders then compete with each other for the item. The highest price wins the auction. Similarly, an item might be requested; and bidders compete to sell that item to the requester at the lowest price that they are willing to sell it. If some bidders have very fast (say millisecond) access to the market, they are in a much better position to bid than are others with much slower access (say seconds). In this case, this is an "unfair" market. It gives an advantage to certain bidders over others.

Contemporary examples of auction markets are Internet auction sites such as eBay, in which a wide variety of objects are put up for sale. These markets are generally manual in that competing bidders are interfacing to the market via their keyboards. Those with faster access can bid more quickly than those with slower connections, leading to an unfair advantage for them.

A more critical case is that of computer trading. If computers are monitoring the market and are automatically making bids, trading activity occurs at a much higher rate – perhaps significantly faster than the difference in access times. An excellent example of computer trading is the stock markets. In these cases, a few millisecond advantage can have a tremendous financial value to the lucky trader.

Figure 4-11 illustrates a hypothetical example in which traders are trading on two different stock or currency exchanges located thousands of miles apart. Consider two hypothetical stock exchanges - the New Jersey Stock Exchange (NJSE) in the USA and the Bosnian Stock Exchange (BSE) in Bosnia. If simple direct access is used (Figure 4-11a), traders may access their local exchange in about 5 milliseconds; but due to the speed of light limitations (20 milliseconds round trip per 1,000 miles over copper or fiber), it takes about 80 milliseconds to access the remote exchange. Thus, New Jersey traders have much faster access to the NJSE; and Bosnian traders have much faster access to the BSE for getting quotes and making trades. This is an unfair system.

Currently, many stock exchanges are affected by regulations that govern the allowable difference in access times across the trading community. How can one even out the access times so that all traders are treated fairly?

An active/active configuration using synchronous replication is one way. This configuration is shown in Figure 4-11b, where synchronous replication may be implemented either via network transactions or coordinated commits. Other implementations might include acquiring a lock or a token on the remote exchange before proceeding.

The underlying principle is that users at each exchange are treated the same regardless of with which exchange they are dealing. Those local to one exchange must wait for their trade to propagate to the other exchange, and vice versa. This configuration has the following active/active advantages:

- Since it is an active/active system, it has extreme availability.

- Since it uses synchronous replication, no trades will be lost in the event of a system failure; and there will be no data collisions.

Moreover, this configuration also enforces fairness:

- Since all queries are done against the local database, monitoring the market is fast and fair. Query access might be in the order of 5 milliseconds.

- Since all trades must be made across the network before they are completed, trading is fair. Network delays may add something in the order of 80 milliseconds to a trade, but this occurs no matter on which end the trader sits.

This is an excellent example of the advantage of coordinated commits over network transactions to accomplish synchronous replication. Because trades will generally involve updates to multiple database entities, the need to send these over the network one at a time (costing 80 milliseconds per round trip) will significantly slow down network transactions relative to the application latency that coordinated commits impose.

Synchronous replication requires that a node must acquire locks on all affected rows across the network before proceeding with row updates. Depending upon how data locks are released, there may be a window in which the new data is available to one side earlier than the other because its locks are released earlier.

There is also the possibility of network deadlocks caused by *lock latency*[63] should both nodes attempt to lock a data item at the same time. Lock latency is caused by the delay in sending the lock status over the replication channel. As a consequence, each trader could obtain a local lock on the same row and then be unable to obtain the lock on the remote row.

In this case, there are several lock resolution strategies that might be employed:

- The trade could be aborted and the traders requested to resubmit the trade.

- Upon receipt of the deadlock notification, each node could wait a random time and then resubmit the trade on behalf of the trader.

- A winner is chosen via some algorithm, and the other party is queued to go next.

- A winner is chosen via some algorithm and the other party backs off and retries at a random later time.

---

[63] See Chapter 3, Building Active/Active Systems.

- The trade is negotiated by some algorithm.

Fairness can also be obtained using asynchronous replication by requiring that all trades be made against the remote database. By doing this, both quotes (local access) and trades (remote access) have equivalent access from either end; and all traders are treated the same. There are no deadlocks, and there is no window of difference due to releasing locks. However, data collisions will occur during the execution of trades and must be resolved in a manner similar to deadlock resolution, discussed above. Remember Rule 32, found in Chapter 9 of Volume 1:

Rule 32: *Lock latency deadlocks under synchronous replication become collisions under asynchronous replication.*

Although these techniques in principle enforce fairness, they depend upon the round-trip time from either end to be essentially the same. However, the round-trip time could be different for each direction for a variety of reasons. For instance, different communication technologies may be used; or transient conditions may slow down communication in one direction.

One solution to this is to specify a minimum round-trip time and hold up transactions that are ready to finish sooner until this time has passed. Another would be to monitor the running average of the round-trip times and autoadjust the imposed delay to provide a self-adapting system as conditions changed.

Network Topologies

The nodes in an active/active system can be arranged in a number of ways. The topologies described below all assume that nodes are connected by dedicated communication channels. The one exception is the last topology which discusses active/active nodes communicating via networks such as the Internet or an intranet.

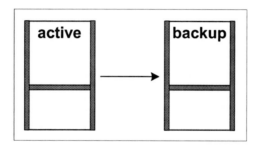

Unidirectional Disaster Recovery
Figure 4-12

Unidirectional Disaster Recovery

We start with the classic active/backup system commonly used today for disaster recovery. Though not an active/active system, it provides a sound basis for the comparison of more advanced architectures. In this configuration, a cold backup (or standby or passive) system is kept synchronized with the active system via unidirectional data replication (Figure 4-12).[64]

---

[64] Although described here as purely passive systems (i.e., not performing active work), many backup systems can be used for read-only activity, such as reporting. Though this improves their usefulness, the active/backup architecture is still not comparable to a truly active/active system.

This configuration falls short of an active/active system for two reasons:

- The backup system is not being used in the application. In effect, 200% of the required capacity must be purchased and only half of it used for the application at any one time.

- Recovery typically takes minutes to hours. This is primarily due to the time required following a failure of the primary to resolve database consistency, to activate the backup applications, and to switch over the users (see Figure 4-13[65]). The extended recovery time precludes the extreme availability that we are looking for in an active/active system.

Why, then, are enterprises still using active/backup systems when it would cost little to reconfigure them as active/active? The reasons are many:

- The developers or IT managers may not yet be knowledgeable in active/active technology.

- In some industries, active/active technology is still new and relatively untried.

- The IT managers may not know or realize that active/active technology is a competitive business advantage.

- Complacency exists among IT managers. They have a unidirectional disaster recovery plan in place and have neither the time nor the resources to investigate alternative strategies.

---

[65] Figure 4-13 is a reproduction of Figure 6-3 in Chapter 6, <u>RPO and RTO</u>, *Breaking the Availability Barrier – Volume 1*, and is described in more detail there.

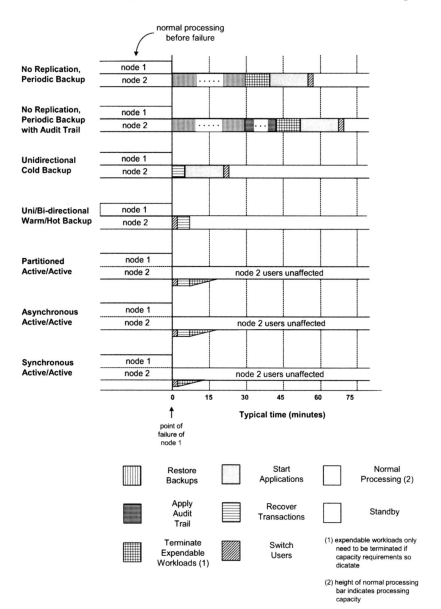

**Typical Recovery Times
Figure 4-13**

- Active/active systems may be perceived as being much more expensive.

- There may be a perception that the application architecture itself precludes an active/active implementation.

- Active/active architectures are often mistakenly thought to be complex to implement and manage.

- Concern may arise about data collisions if asynchronous replication is used.

- There may be a concern about the extended response time due to application latency if synchronous replication is used.

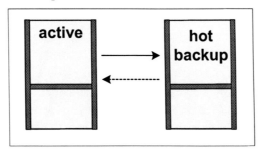

Sizzling Hot Takeover
Figure 4-14

Sizzling Hot Takeover

In this configuration (Figure 4-14), two nodes are ready to process transactions as if they were in an active/active configuration. Their databases are synchronized via data replication. However, only one node is actively processing transactions while the other node acts as a hot backup. The backup node's database is kept synchronized with the active database via replication.

Though the backup system is not carrying any update transaction load (though it could be used for read-only application purposes), it is otherwise prepared to accept transactions. All applications are up and running, and all database tables and files have been opened. One can even periodically submit test or verification transactions to the backup applications in the hot backup node to verify that it is functioning properly.

Consequently, should the primary system fail, the backup system is immediately ready for users to be switched to it; and recovery from a primary failure can be done in seconds.

This characteristic provides a very important capability. All recovery plans should be periodically executed to make sure that no configuration or other changes have occurred which would make the plan inoperable. It is also important to keep the data center staff current in failover procedures. Recovery testing is often a risky task as the failover procedure may fail and take the entire system down. With a sizzling hot configuration, it is always known that the backup system is immediately capable of taking over from the active system. This can be verified simply by sending a test or verification transaction to the backup system. Therefore, failover can be practiced frequently with very little risk. In fact, if the capabilities of the active system and backup system are equivalent, the backup system could become the new active system following a failover test.

This configuration, of course, is not truly an active/active system since not all (in fact, only half) of the purchased processing capacity is being used. However, its very fast recovery time leads to the extreme availability characteristic of active/active systems.

Sizzling hot takeover is an option for those applications that for one reason or another cannot run fully active/active but yet require instant failover. For example, the application architecture may require that there be only one active application instance; and the application or replication engine cannot be modified to resolve this limitation. A process control application that must process all events in strict time sequence is such an example.

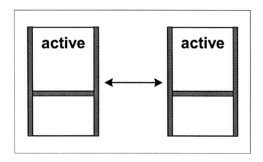

Bidirectional Active/Active
Figure 4-15

Bidirectional Active/Active

The simplest active/active architecture is that of two nodes synchronizing via data replication (Figure 4-15). In this case, all available processing capacity is used; and recovery from a failure is measured in seconds.

This architecture is often a first step in the migration from an active/backup architecture to an active/active architecture. However, as discussed previously, running in an active/active mode introduces the data collision potential, which must be addressed. The rest of the architectures discussed below assume that an active/active architecture is in use.

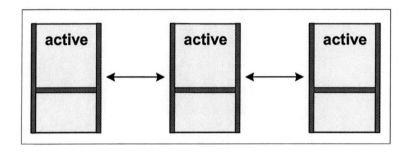

Route Thru
Figure 4-16

Route Thru

If there are multiple application nodes which are significantly distributed geographically, then the nodes need not be fully connected. Some active nodes may also be routing nodes (Figure 4-16).

For instance, if nodes are located in London, San Francisco, and Sydney, changes made in London may be routed to Sydney via San Francisco and vice versa. This simply means that changes received by San Francisco from London update the San Francisco database and then are replicated to Sydney via normal data replication. Likewise, changes from Sydney are replicated to San Francisco and then rereplicated to London.

Note that this is a simple example of a routing table indicating where to rereplicate changes received from another node, as more fully described in the next section.

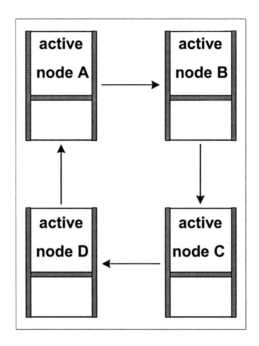

Ring
Figure 4-17a

Ring

An active/active ring network is shown in Figure 4-17a for four nodes. In this configuration, updates made by one node are rereplicated around the ring to all other nodes in a manner similar to that described previously for the route-thru configuration. However, in this case, the routing tables must be set up to avoid the continuous circulation of a change around the ring.

As shown in Figure 4-17a, each node replicates its changes to its immediate clockwise neighbor. It furthermore acts as a relay node to pass changes from other nodes, excluding its immediate clockwise

195

node, to its clockwise node. For example, node A will pass changes made by nodes C and D to node B. However, it will not pass changes made by node B to node B in order to avoid continuous circulation. In this way, a change made to one database copy is replicated to all other database copies in the network.

In addition, alternate routing rules must be defined to reroute around a node which has failed.

Replication routing tables are similar to network routing tables except that they are more complex. Network routing tables generally define the next hop in the network to which to deliver a message in order to keep it moving to its intended destination, and they are therefore destination-based. That is, they define to which node a message should be routed based on the message's *destination*. Replication routing tables must route to multiple nodes in the network (in effect, a multicast). In addition, they must have the intelligence to prevent loops. This means that the nodes to which a message is sent depends upon from which node the message was received.

As an example of a routing table, and with reference to Figure 4-17a, the following routing table might be defined for node A.

Source	Send To
Node A	Node B
Node B	X
Node C	Node B
Node D	Node B

Primary Routing Table for Node A

In this table, the "source" is the node at which the change originated. Each node replicates its changes to its immediate clockwise neighbor. In addition, each node rereplicates changes received from other nodes, excluding its clockwise neighbor, to its clockwise neighbor. For instance, node D's changes are rereplicated to node B by node A as are node C's changes. However, node B's changes end their transit at node A and are not rereplicated. Similar routing tables are established for the other nodes. Routing for node A is illustrated in Figure 4-17b.

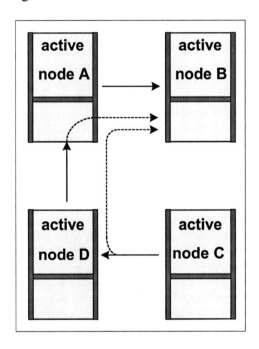

Node A Routing
Figure 4-17b

Should a node fail, its neighbors cannot replicate to it. Rather, changes are queued and sent to the node when it returns to service. Since the failed node can no longer function as a route-thru node, the direction of replication must be modified. In the example of Figure 4-17b, nodes A and D normally receives changes from node B via node C. Should node C fail, node A must now serve as the relay node from node B to node D. Node A's routing table now becomes

Source	Send To
Node A	Node B
	Node D
Node B	Node D
Node C	none recvd
Node D	Node B

Routing Table for Node A
Node C Failure

Node A replicates changes which it makes to both of its neighbors, nodes B and D. Node B's changes are replicated to node D and vice versa via Node A. Node B queues changes from nodes A, B and D to send to node C when it is restored to service.

Similarly, special node A routing tables must be established for node B or D failures.

Note that because a node failure will require reverse replication at some intermediate node, each replication path must comprise a bidirectional replication engine even though normal replication is in just one direction.

One problem with the above routing is that the node on the far end of the chain must wait for three replication intervals to receive changes from the node at the beginning of the chain. For instance, node D must wait for changes from node A to propagate through nodes B and C. This increases replication latency and the possibility of data collisions and data loss following a failure.

This situation can be improved by taking advantage of bidirectional replication and having each node replicate to its immediate neighbors. With this strategy, node A's routing is as shown in Figure 4-17c. Node A will replicate its changes to nodes B and D, and will relay node D's changes to node B.

The routing table for node A for this case is as follows:

Source	Send To
Node A	Node B
	Node D
Node B	X
Node C	X
Node D	Node B

Primary Routing Table for Node A
Bidirectional Replication

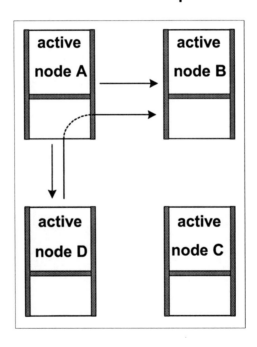

Bidirectional Node Routing
Figure 4-17c

Should node C fail, node A takes over its relay function via the same routing table defined above for a node C failure.

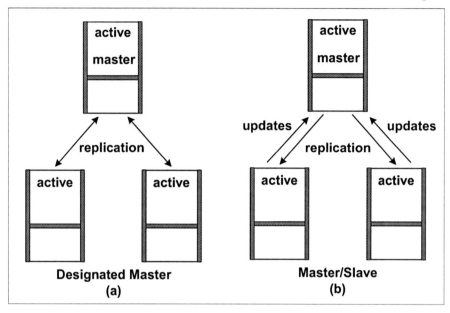

Hierarchical
Figure 4-18

Hierarchical

In a hierarchical topology (Figure 4-18), one node acts as a master and is responsible for routing all changes to the other (slave) nodes. We consider two cases of interest.

The first case is that of a *Designated Master*. One node in the network is designated as the master node. The other nodes are known as slave nodes. Each node in the network applies its changes to its local copy of the database. The slave nodes replicate their changes to the master node, which then replicates them to all slave nodes in the network including the originating node.

It is quite possible that the master may get colliding changes from two or more of its slave nodes. In this case, any of the collision resolution techniques described earlier may be used to resolve these data collisions at the master. Since the master rereplicates all changes to all nodes including the originating nodes, all nodes are guaranteed to converge to the same data state.

The second case is the *Master/Slave* configuration. In this case, one node is again designated the master node. However, instead of the slave nodes updating their local copy of the database and replicating those changes to the master node, they make their changes directly to the database copy resident on the master node. The master node then replicates those changes to all slave nodes. Because all changes are made only to the database on the master node, there will be no collisions.

These hierarchal configurations solve the database of record problem in active/active systems, as the master's database is the database of record.

One issue with the hierarchical topology is that there is only one master node. Should that node fail, then the network is left without a master node. The solution is to promote one of the slave nodes to become the new master node until the master node is returned to service. Promotion is accomplished by simply modifying the routing tables in the nodes.

In this topology, the master node might also be considered a hub node since all traffic passes through it.

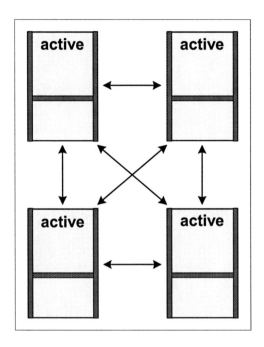

Fully Connected
Figure 4-19

Fully Connected

In a fully connected network as shown in Figure 4-19, each node is connected to all other nodes. In this case, routing is trivial since each node sends its changes to all other nodes. Routing tables are not required.

Should a node fail, then changes being replicated to that node simply queue in the replication engines feeding it from the other nodes. When the node is restored to service, these queues drain, replicating their changes to the failed node in order to resynchronize its database.

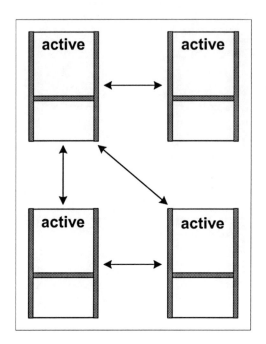

Partially Connected
Figure 4-20

Partially Connected

A partially connected topology (shown in Figure 4-20) is similar to a fully connected network except that not all nodes are directly connected. Route-Thru and Ring topologies are special examples of partially connected networks.

Partially connected networks require dynamic routing tables to instruct nodes how to rereplicate changes between nonconnected nodes. Additional routing tables must be defined to reconfigure the network in the event of a node failure, or some other form of route discovery algorithm may be used.

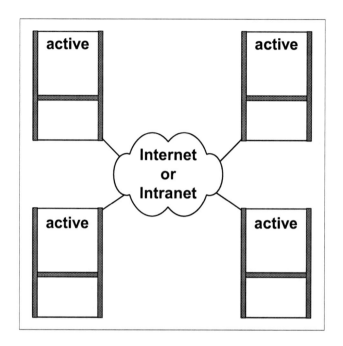

Internet/Intranet
Figure 4-21

Internet/Intranet

Rather than being interconnected by dedicated communication channels, the nodes in an active/active network may communicate with each other over a network such as the public Internet or a private intranet. Thus, they communicate via a fully-connected network since any node can communicate directly with any other node.

If there are only a few nodes in the application network, then a node might establish a dedicated TCP/IP connection with each of the other nodes. This is equivalent to the fully connected topology shown in Figure 4-18.

However, this topology offers an improved way to communicate if the application network is large; that is, if there are many nodes in the active/active network. Since every node needs to send its database updates to all other nodes, this is fundamentally a multicast problem. Most networks such as a company's intranet support broadcasting.

In the case of the Internet or an intranet, a node would register with all other nodes from which it wishes to receive database updates. It would then send changes to all nodes which had registered with it.

A simple UDP broadcast is not sufficient since there is no guarantee that an update will be delivered. Therefore, a layer must be added on top of UDP to provide a reliable broadcast method. For instance, each change block could be sequentially numbered. If a node detects that it is missing a block, it could rerequest that block from the sending node via a reverse TCP/IP connection.

Also, one must be careful with UDP as it rarely works across subnets, and it may cause a quality-of-service problem since UDP packets are given the lowest priority.

Again, if a node should go down, the other nodes in the application network must queue their database changes to that node and deliver them when the node once again returns to service.

What's Next

The network topologies above all require reliable network connections for internodal traffic. This traffic not only includes replicated data but also heartbeats to monitor node health and user traffic if a group of users has to be switched from a downed node to an operating node.

Various ways to configure and manage these networks are described in the next chapter, Redundant Reliable Networks.

Chapter 5 - Redundant Reliable Networks

"Keep thy shop and thy shop will keep thee."
- Benjamin Franklin

The Need for Network Redundancy

Networks are the nerve fibers of active/active systems. They tie the parts together and allow those parts to coordinate. Lose a network, and the system may flail or fail. Therefore, it is important that each of the many networks used throughout an active/active system be just as reliable as its nodes and databases.

As we have shown, nodes and databases achieve their reliability through redundancy. So, too, for networks. For every path from one critical component in the system to another, there must be an alternate path to use in the event of a path failure.

We have studied the availability relationships extensively in Volume 1 of this series,[66] and the results are summarized in Chapter 2, <u>Reliability of Distributed Computing Systems</u>, of this volume. Two results apply directly to networks:

- Providing a redundant network doubles the 9s of a single network. If a network (for instance, a LAN or a WAN line) has an availability of four 9s, then providing a backup will provide an availability of eight 9s for the connection

---

[66] W. H. Highleyman, Paul J. Holenstein, Bruce D. Holenstein, *Breaking the Availability Barrier: Survivable Systems for Enterprise Computing*, AuthorHouse; 2004.

- Two smaller networks are better than one larger network [see Equation (2-27)]. This applies especially to breaking up large LANs into smaller subnets.

These observations lead to two network corollaries of previous rules:

Rule 51: (Corollary to Rule 2) - *Providing a redundant network doubles the 9s of the network.*

Rule 52: (Corollary to Rule 9) - *Two smaller networks are more reliable than one large network.*

In this chapter, we will explore the means for configuring redundant networks, for providing fault detection within these networks, and for failing over to alternate paths in the event of a link or network failure.

Networking is a massive field in itself. We will discuss various approaches to providing redundancy in networks, but the details are left to other volumes on this subject, as referenced.

Reliability Is More Than Just Redundancy

The availability of an active/active system is dependent upon the availability of each of its parts, primarily its nodes, its databases, and its networks. But high network availability requires much more than just redundant paths.

Reconfiguration Time

The time to reconfigure the network is the mean-time-to-repair (MTR) component of the availability equation. Reconfiguration of networks requires several steps:

- determining that there is a problem
- determining the source of the problem
- determining the solution
- implementing the solution

Depending upon the complexity of the network and the technologies in use, each of these might be an automatic process or a manual process. In the ideal case, networks should fail over automatically and lead to MTRs measured in seconds. However, for very complex networks, automatic failover may be too complicated and therefore deemed to be too risky. Manual failover could take minutes to hours.

Therefore, the choice of the network technology to be used for a redundant network goes far beyond cost. The means for network monitoring, fault detection, and failover strategy must all be carefully considered when making the choice.

Capacity

Reconfiguring around a failed link may generate additional load on an already active link or may make use of a backup link with different throughput capabilities (such as a dial-up link).

It is important to ensure that the network still has the capacity to carry all critical traffic in the event of a failure. Otherwise, the reconfigured network itself may shut down due to overutilization. Ethernet LANs are particularly susceptible to this effect. Because of contention caused by multiple sources trying to transmit simultaneously over the LAN, such a LAN can only be loaded somewhere between 10% and 40%, of its maximum capacity depending upon the length of the LAN.[67] Depending upon the

---

[67] W. H. Highleyman, Chapter 5, <u>Communications</u>, *Performance Analysis of Transaction Processing Systems*, Prentice-Hall; 1989.

networks in use, load shedding (i.e., shutting down noncritical applications) may be necessary in the face of a network failure to prevent network overloading.

An extreme real-life example of network overload, though it was a power network and not a communication network, was the great Northeast Blackout of 2003 in the U.S. and Canada. The failure of one transmission line cascaded into heavy loads on other components. Consequent consecutive failures occurred until the entire power grid serving this area went down.

Latency

Following a network failover, data traffic will be rerouted over another link. If the latency of this link (i.e., the time that it takes for a signal to traverse it) is greater than that of the original link, then response time will suffer.

The latency of a link is not only affected by its distance (the speed of light through fiber or copper is about 10 microseconds per mile), but also by each router, switch, bridge, hub, or other equipment in its path. Even worse, the latency through shared networks such as the Internet can be indeterminate.

Therefore, one must ensure that the anticipated latency of the backup channels is such that the response time to users remains acceptable.

Security

Based on the users' needs, each network has a security policy and must provide the requisite protection against snooping, spoofing, attacks, and other outside interferences. These security policies must be enforced by the backup channels as well.

The Great Protocol Wars of the Twentieth Century

The protocol wars that were waged into the late 20th century are over, and the winner for now is IP (Internet Protocol). Though not relegated to the dustbin, contenders such as X.25 and SNA have become niche players.

IP is such a pervasive protocol that nearly all modern enterprise systems depend upon it for intercommunication. Therefore, we will consider only IP networks, be they LANs or WANs.

Redundancy Configurations

There are two classes of redundancy configurations that we will explore:

- *backup networks*, which provide *multiple networks* for failover. These could be used in active/backup configurations or for load sharing when both are operational. Dual LANs and dual WANs are examples of these.

- *reconfigurable networks*, which automatically reroute around a fault in the network. Packet switches. such as those used in the Internet backbone, are examples of these.

There are cases in which these two configurations may be mixed to provide the proper degree of redundancy.

Backup Networks

A backup network is a separate network from the primary network and is usually dedicated to providing the same interconnectivity that the primary network provides. It must be totally independent of the primary network so that there is no single point of failure that could take down both networks. For instance, the two network rails should be connected to nodes or user consoles via separate network interface cards (NICs) or communication adapters.

Backup networks are applicable to both local networks and long-haul networks, as discussed below. They may be used as a primary/backup pair or in a load-sharing configuration.

Local Networks

Local networks are typically LANs (local area networks) that are used to connect users to each other or to their home node. A LAN may also be used for internode communications for collocated nodes.

Dual LANs

A typical redundant LAN connecting a community of users to their home hosts is shown in Figure 5-1a. Two independent LANs are provided. The host and each user console connects to each LAN via independent NICs.

In this configuration, one LAN can be the active primary LAN; and the other can be the passive backup LAN. In the event of a failure of the primary LAN, all traffic is rerouted to the backup LAN, which now becomes primary. When the failed LAN is returned to service, it becomes the new backup.

Alternatively, both LANs could be active in normal operation (for instance, by routing requests in a round-robin fashion). Should one LAN fail, the surviving LAN carries (and must have the capacity for) the entire load until the failed LAN can be returned to service.

Finally, both LANs could be carrying all traffic, with the host discarding duplicates.[68] Should one LAN fail, there is no failover procedure required. The host continues to receive all traffic over the surviving LAN.

Virtual IP Address

The simple configuration described in Figure 5-1a suffers certain deficiencies:

- How does a console know which LAN is the currently active LAN?

- How do users get switched over to a backup host in the event that their home host should fail?

These problems are solved by using routers to implement a virtual IP address, as shown in Figure 5-1b.[69] In this configuration, each console knows about its host only via a common IP address (shown as IP0) which never changes. All of its requests are directed to address IP0. The console can use either LAN to send requests to one of the redundant routers via the address IP0.

Meanwhile, the host has advertised to the routers to which of its two interfaces IP0 traffic is to be routed. If the path to the currently active interface goes down, then the surviving interface advertises that it is now the recipient of IP0 traffic.

---

[68] For instance, the host's protocol stack both sends and receives packets on both rails. Duplicate packets and rail failures are handled transparently to the application.
[69] E. Marcus, H. Stern, Chapter 8, Redundant Network Services, *Blueprints for High Availability*, Robert Ibsen; 2000.

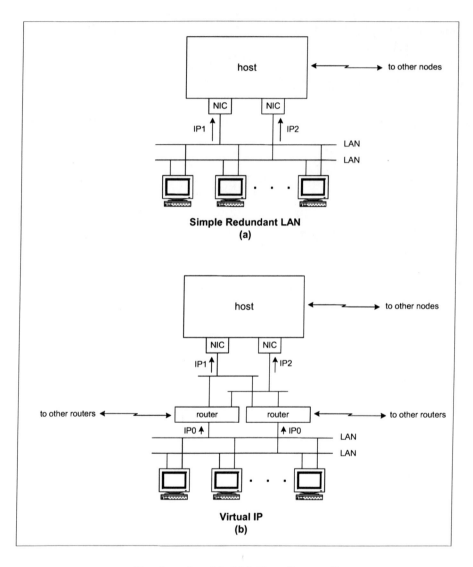

Redundant LAN Configuration
Figure 5-1

Likewise, a console that finds it cannot communicate over one LAN can switch to the other LAN. The router on that LAN will route its traffic to the appropriate node interface.

Should the host fail, a backup host can similarly seize the address IP0 by advertising to the failed host's routers that IP0 traffic should now be directed to it.

The detection of host failures and the switching of users to surviving hosts is discussed in more detail in Chapter 7, <u>Node Failures</u>. As described previously, this configuration can also be used in a primary/backup mode, in a load sharing mode, or in a simultaneous parallel mode.

Long-Haul Networks

Long-haul networks, or WANs (wide area networks), are required whenever the interhost distance or user-host distance is greater than the capabilities of a LAN. Long-haul links may be privately owned (such as microwave links) but are more commonly implemented as a service from a public carrier. Such a service may be leased lines, T1 or T3 links, OC-1 or OC-3 links, DSL links, satellite links, ATM services, or others.

Leased lines and T1/T3 links are typically dedicated channels with no interference from other traffic. Other types of channels may compete with the traffic of other users. This may lead to poor performance during busy times. It is important to have service level agreements (SLAs) in place for such shared facilities.

There are several options for backing up a long-haul circuit:

- The link can be replicated using a similar link with the same performance (or even lesser performance if reduced capacity is deemed acceptable). However, care should be taken to use a different network provider or technology.

- A dial-up connection can be used, again if its capacity is deemed acceptable.

- The Internet can be used as the backup channel. This can be very inexpensive and is normally very fast. However, the Internet can become heavily loaded during peak hours, with commensurate poor performance.

As with LANs, the connections to the primary and backup channels must be independent, with no single point of failure. At the very least, this means that interfaces to these channels must be through different communication adapters. Furthermore, the channels should be obtained through different communication carriers to avoid an outage due to a carrier problem. It is also advisable to ensure that the two carriers involved do not share the services of a third carrier, the failure of which could take down both networks.

In addition to ensuring that the capacity and reliability of the backup channel is sufficient, it is important to ensure that the company's security policy can be enforced with the backup channel. An Internet link, even when tunneling protocols such as VPN are used, will be much less secure than a private dedicated link.

Reconfigurable Networks

Network redundancy can also be provided by complex "self repairing" networks such as telephone backbone networks and packet switching networks. These methods provide multiple paths with automatic rerouting around failed paths or failed nodes or even in some cases around overloaded paths or nodes.

Local IP Networks

In large campus or metropolitan clusters, LAN networks can be quite complex. They comprise multiple subnetworks interconnected by routers, switches, bridges, and gateways.[70] A router may be a physical router device or a host configured as a router connecting to multiple subnets (this is called "multihomed").

Alternate Routes

In local IP networks, there typically is any number of routes to be found that can be used to carry traffic between any two given points.[71] Some routes may be direct connections between the two end points; others may require transit through one or more routers. Passage through a router is called a *hop*.

For instance, Figure 5-2 shows three subnets – Subnet 1, Subnet 2, and Subnet 3. Router 1 connects Subnets 1 and 2. Router 2 connects Subnets 2 and 3. The multihomed Host 1 is also configured as a router connecting Subnets 1 and 3. Host 2 is directly connected to Subnet 1, and Host 3 is directly connected to Subnet 3.

If Host 1 and Host 2 need to talk, they can do so directly through Subnet 1, to which they are both connected (zero hops). However, if Host 1 should lose its connection to Subnet 1, it can still reach Host 2 via the route Subnet 3 to Router 2 to Subnet 2 to Router 1 to Subnet 1 to Host 2 (two hops).

---

[70] John McNamara, *Technical Aspects of Data Communication*, Digital Press; 1988.
[71] W. Stevens, Chapter 9, IP Routing, *TCP/IP Illustrated, Volume1 – The Protocols*, Addison-Wesley; 1994.

Likewise, Host 3 can talk to Host 2 via Router 3 (one hop). Should Host 1 go down, thus taking down Router 3, Host 3 can still reach Host 2 via Router 2 and Router 1 (two hops).

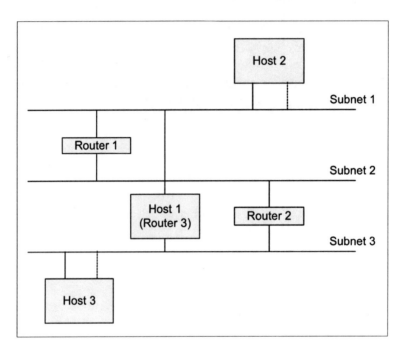

Simple Local Network
Figure 5-2

In setting up a router, it is imperative to ensure that there are no loops which will cause a message to move through the network indefinitely and never get delivered. Such a packet may eventually get dropped if it has exceeded its TTL (time-to-live) parameter.

Dynamic Routing

All of the routers maintain a current map of the entire network. This is done via the Routing Information Protocol (RIP) or an equivalent protocol.[72] With RIP, each router periodically reports its routing table to its directly connected routers. Thus, each router can see that it can access all of the destinations served by the reporting router with one more hop through that router. As a consequence, each router will build a routing table that may have several routes to a given destination, each with a different *cost* specified as the number of hops required to reach that destination. Each route is specified only as the next router to which to send the message. That router will then pass it on as appropriate to another router or to the destination.

When a router receives a message to be forwarded, it searches its routing table for the least cost route (least number of hops) and passes the message to the specified router. Some extended router protocols provide multiple choices for cost measurement (called metrics) in addition to hops and take into consideration such attributes as throughput, round-trip time, or route reliability. If multiple routes with the same cost are found, the router may optionally route traffic over all of these routes to provide load sharing.

This process is known as *dynamic routing*. Dynamic routing is a powerful mechanism to repair a network following a node or path failure. Routers will typically exchange routing tables whenever a change is made (for instance, a link goes down or a cost changes) or periodically (typically every 30 seconds). It can take several minutes for a network reconfiguration to propagate through the network.

Fragmentation

A message being sent over IP is called a *datagram*. Each IP link has a maximum datagram size that it can carry (512 bytes or larger, with a further Ethernet limit of about 1500 bytes). The maximum datagram size that can be sent over a route is that of the link with the

---

[72] W. Stevens, Chapter 10, <u>Dynamic Routing Protocols</u>, *TCP/IP Illustrated, Volume1 – The Protocols*, Addison-Wesley; 1994.

221

smallest size datagram capability. This is known as the maximum transmission unit (MTU) of the route, and a sender can query a route for its MTU.

If a datagram that is larger than the MTU for the next link is received by a router, then it must *fragment* the datagram into packets.[73] Fragmentation can occur at multiple points along a route should path MTUs become smaller.

Once a datagram is fragmented, each of its packets is sent individually as an IP message with no knowledge of the other packets. This means that each packet could be sent over a different route and that therefore the datagram packets could arrive at the receiver out of order. Some packets could even be dropped and have to be rerequested. It is up to the receiver to rerequest and reorder the packets to recreate the original datagram.

Fragmentation is a performance issue since each packet must individually carry an IP header (currently, 20 to 60 bytes in IPv4, depending upon the number of IP addresses in the header).

Long-Haul IP Networks

IP Networks

Figure 5-2 shows a locally connected LAN network. There is no reason why such a network could not be extended over long-haul networks. Figure 5-3 shows this same network geographically distributed. Geographic distribution is generally needed for redundant nodes to provide disaster tolerance and/or data locality.

---

[73] W. Stevens, Chapter 2, <u>Link Layer</u>, *TCP/IP Illustrated, Volume1 – The Protocols*, Addison-Wesley; 1994.

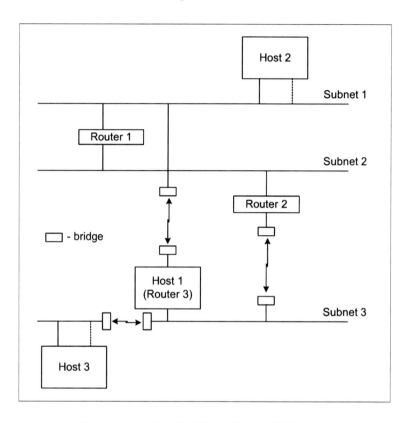

Geographically Distributed Network
Figure 5-3

In this figure, Subnet 3 is geographically a long distance from Subnets 1 and 2. Router 2 connects to Subnet 3 via a long distance channel, and Host 1/Router 3 connects to Subnet 1 via a long distance channel. Likewise, Subnet 3 is extended over a long distance to Host 3. Connections to the long-haul channels may be done via repeaters, switches, bridges, or routers.[74]

---

[74] Not shown are firewalls which are generally needed when connecting to public networks. If firewalls are used, they too should be redundant.

223

The bridging between routers and networks makes this geographical extension transparent to the IP process. Routing and dynamic routing work much as described above.

Of course, in this case, the number of hops may not be a good cost metric. It may be faster to go through five local routers than it is to go through two remotely connected routers because of latency and throughput restrictions of the long-haul channels. This is the importance of extended cost metrics, such as the round-trip time provided by some routers when making routing decisions.

The above discussion has focused on private networks. However, these same methodologies can be extended to public packet switching IP networks and are, loosely speaking, the foundation of the Internet backbone.

Telephone Networks

Though not related technically, telephone networks share the same characteristic of dynamic routing as evidenced in IP networks. This is especially true for modern services such as ATM (asynchronous transfer mode), which will route around a failure transparently to the user. Should a failure occur, traffic is often rerouted without dropping a session or connection.

Dialed Backup

Dialed backup is a mix of backup and alternate routing concepts. It can be used to establish a new route through the network if a primary route should fail. Of course, with this sort of redundancy, a primary route failure will result in connection loss until a backup connection is established via the dialed backup. Likewise, the failure of a route assigned to a dialed connection will also cause a connection loss; and the connection will have to be redialed (giving the network a chance to find another route).

Backup Connections

Even though we are depending upon alternate routing through the network, it must be ensured that each host has a redundant connection to the network. In Figures 5-2 and 5-3, Host 1 has two connections, one to Subnet 1 and one to Subnet 3, and therefore can reach any end point in the network through either connection.

However, Hosts 2 and 3 connect to only one subnet. Therefore, each should be provided with two independent interfaces to their local subnet.

In addition, to meet the full redundancy criterion, Subnets 1 and 3 should be redundant, as shown in Figure 5-1b, since these subnets represent a single point of failure for Hosts 2 and 3, respectfully.

The Internet

Using the Internet to provide component connectivity is an important consideration. The Internet has several advantages for this purpose:

- It is highly robust. It will automatically route around failures and deliver traffic reliably (just be sure to have dual independent interfaces to the Internet as well as dual service providers.)

- It is fast. Service in the megabits/second range can be provided.

- It is cheap. Monthly ISP (Internet service provider) costs are a small fraction of the cost of a leased service.

However, there are also concerns that should be addressed.:

- Its response time is not controllable. During peak traffic times, delivery of messages can be significantly delayed (more likely the ISP than the Internet backbone); or messages may be dropped and will have to be resent.

- Its security level may be insufficient to satisfy the corporate security policy.

The Internet can often be a serious candidate for internodal channels – especially for backup purposes. Just be careful of its frailties.

An extension of the Internet is the corporate Intranet. This is a network using the same technology as the Internet but that is privately owned and managed. Therefore, issues such as peak loads, response times, and security are more under the control of the corporate owner.

In addition, Virtual Private Networks (VPNs) can be established over either the Internet or an Intranet to isolate a community of users even more.

Fault Detection

The first step in fault recovery is to detect a fault. Only then can action be taken to correct the fault. There are several techniques that can be used to monitor a network connection for faults.

Heartbeats

Heartbeats are a very effective network monitoring mechanism. In effect, one end of a connection will send a periodic message to the other end to ensure that there is an operable path. In an IP network, heartbeat messages can be generated by the application by sending *pings*, an IP facility. For non-IP links, the heartbeat is a simple message. However, there are problems with pings. Firstly, they may be blocked by firewalls. Secondly, they verify only that the network is up but say nothing about the applications. End-to-end heartbeat messages are therefore more effective.

IP often generates *keep-alive* messages to verify that the other side of a connection is still there. However, keep-alive messages are generally not a suitable heartbeat as their interval is too long. Also, some implementations of IP do not support keep-alive messages.

Should a heartbeat be lost, one should not immediately conclude that the path is down. The heartbeat message may have been undeliverable for other reasons, such as corruption due to line noise. However, if several heartbeats in a row are missed, then it can be concluded that the path is down.

Typically, three heartbeats must be missed to declare a failure. For instance, if heartbeats are sent every three seconds, a path failure will be detected within nine seconds. In many systems, data messages substitute for heartbeats. A heartbeat message is sent only if a data message has not been sent for a while.

Heartbeats bring with them a performance/recovery time compromise. As the heartbeat rate is increased, failure detection time (and thus the recovery time MTR) is reduced; but network load is increased. This may not be a governing factor for one client pinging one host but certainly will be for hundreds or thousands of clients pinging multiple hosts in large networks.

Heartbeats are generally sent by clients to their servers of interest. It often is not feasible for hosts to ping their clients since they do not necessarily know who their current clients are. Servers are expected to be persistent; clients are not.

Heartbeats may be of two types:

a) *Passive* – A heartbeat only is sent. There is no expectation of a reply. Therefore, only the receiving end can detect a failure. This is not suitable for client/server heartbeats but may be useful for host/host "I'm alive" messages in which two or more hosts in a network are monitoring the health of others (see Chapter 6, Node Failures).

b) *Active* – A heartbeat message requires a response. In this way, both ends can detect path viability. This is required for client/server heartbeats and can be useful for host/host heartbeats.

It can be useful to number heartbeats sequentially. Missing heartbeats can then be tabulated and used as a measure of path reliability. If a path starts to miss heartbeats at a significant rate without a path failure, this is a sign that the path may be becoming unreliable and that an alternate path should be considered.

Missing Responses

Client requests can be substituted for heartbeats. In effect, a path failure will not be noticed by a client until it tries to use it, at which time it will try several times and then switch to an alternate path.

Analogous to heartbeats, the absence of a response to a client request is an indication of path failure. After a certain set of timeouts and retries (say three), either the path or the target system can be declared down.

Client requests and heartbeats can also be mixed. A heartbeat is sent only if a request has not been sent within the heartbeat time.

TCP Detection

IP is a connectionless protocol. Packets are simply sent, and it is assumed that they will reach their destination (though they may not). To receive confirmation that a message has been received, a two-way connection must be established between the sender and the receiver. In an IP network, clients establish connections with their servers via the Transmission Control Protocol (TCP).[75] TCP guarantees the delivery and proper order of all packets.

With TCP, the client application is often unaware of IP addresses. It simply needs to know the name of the service to which it is trying to connect. Under the covers, a host table or a domain name service (DNS) is queried to map the service name to a particular host (IP address) and service running on that host (port number). A connection is established between the client and that service for the duration of the client's needs.

TCP can detect the loss of the session through keep-alive messages. Should a session be lost, the client can reconnect. However, this facility is not suitable for active clients as it can take too long. Typical default times are two hours, but this time can generally be reduced to subminutes. Hence, a client/server end-to-end heartbeat is often implemented.

Path Monitoring

IP networks can provide a range of metrics concerning the current state of a path or network. These metrics are useful for determining path failures or pending path failures. Metrics include, for instance,

---

[75] W. Stevens, Chapter 17, <u>Transmission Control Protocol</u>, *TCP/IP Illustrated, Volume1 – The Protocols*, Addison-Wesley; 1994.

missing packet counts. In addition, sequence-numbered heartbeats can provide a measure of the overall connection reliability.

If packets are not received for a while, there is a good chance that the path has failed. If heartbeats have not been received for a while, it can be safely deduced that the path or sender is down. If there is a high incidence of missing packets or heartbeats, the path's reliability is being compromised. In any event, failover is to be seriously considered.

Fault Recovery

The procedure for recovering from a path fault depends a great deal upon the redundancy configuration of the path.

Alternate Routing

If the path is through a network that can automatically reconfigure itself around link failures, such as IP or ATM networks, then there is no failure mechanism required of clients or servers. Recovery by rerouting is performed transparently to the end points of a connection.

Backup Networks

If dual networks are provided, such as those shown in Figure 5-1, then there are several ways in which they may be used.

Redundant Transmission

One configuration is that both networks carry all traffic. Servers must then listen on each of their redundant interfaces and discard duplicate datagrams. Should a network (or interface) fail, then traffic will continue to be delivered over the surviving connection. No failure recovery is required.

Of course, manual or automatic means must be provided to notify end points that the failed network has been returned to service so data can once again flow over it. For instance, each client can periodically attempt to send requests over the failed network. If a transmission is successful, then it knows the network has been returned to service. Servers will know this when they begin receiving traffic from the failed network.

Load Sharing

Load sharing is similar to the redundant use of dual networks except that traffic is divided between the two. No duplicate traffic is sent.

Should a network fail, the client will either receive no response to a request or find its messages rejected. In this case, it should route all traffic over the surviving network.

As with redundant transmission, a client can detect the return to service of a failed network by periodically attempting to send a heartbeat. (Depending upon a request message in this case may not be acceptable because it may take several seconds for the request to be rejected, thus resulting in an excessive response time to that request.)

Of course, when using load sharing, it is important that one network can handle the full network load or an agreed-upon lesser load if load shedding is used in the event of a network failure.

Primary/Backup

In a primary/backup configuration, all traffic is sent over the primary network. The backup network is normally passive and may not even be connected (such as a dial-up line) until a failure occurs. However, if automatic failover is desired, all servers should be listening on both networks. Otherwise, in the event of a primary network failure, a server will be isolated from any traffic; and failover will have to be manual.

The failure recovery mechanism is different for clients and servers.

Clients: If a client cannot issue a request or does not receive a reply to an outstanding request in an appropriate period of time (and the rerequest count is exhausted), then it should switch its request traffic to its backup interface connected to the backup network. Provided that the server is listening on both networks, then the requests sent by the client over the backup network will be received and processed by the server. Responses will likewise be routed over the backup network.

All requests and responses should carry sequence numbers so that duplicate and missing messages can be detected.

Servers: If a server cannot receive traffic over the primary network, and if it is listening to both networks, then it need take no failure recovery action since all clients will ultimately switch to the backup network.

Virtual IP

In some implementations, it may not be feasible for redundant servers to listen on both networks. For instance, if the server is a cluster (Figure 5-4), in which the primary and backup servers are different systems, database contention restrictions may dictate that only one server can be active at any one time. In this case, the use of a virtual IP address can be useful.

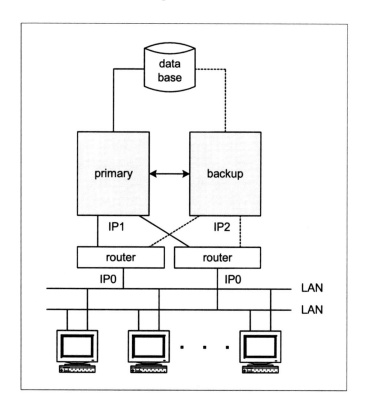

Clustered Server
Figure 5-4

With virtual IP, the clients have no knowledge to which server they are actually talking. They see the server cluster as a single system with a single IP address (IP0 in Figure 5-4).

Each client may still be connected into the network via a dual LAN and may use these in any of the ways described previously (redundant transmission, load sharing, primary/backup). Each LAN is serviced by a router, which can route messages to either side of the cluster.

The routers receive all requests with a virtual IP address of IP0. Their routing tables each point IP0 traffic to the primary side of the cluster, mapping IP0 to IP1 in Figure 5-4.

Should the primary fail, then the backup takes over. The routing tables in the routers must now be changed to route IP0 traffic to the backup's IP address (IP2 in Figure 5-4). This is accomplished by the backup advertising to each router that it is now handling IP0 traffic. Thereafter, the routers will route IP0 traffic to the backup server at address IP2.

An alternate implementation that is sometimes used in clustering products is to send requests to both servers. The servers exchange control information on a private LAN to determine which server will process the request.

Node Failure

So far, we have assumed that the network is at fault if clients cannot talk to their servers. However, what if it is not a network failure? What if a node fails? What if a node loses access to its database?

In the case of a failed node, there is no recourse but to switch all users serviced by that node to designated surviving backup nodes. This is discussed in detail in Chapter 7, Node Failures.

The second case, that of the loss of a database, requires either that the node be able to connect to a remote database or that all users be switched to backup nodes with database access. This case is discussed in Chapter 6, Distributed Databases.

Automatic or Manual Failover

So far we have described ways in which recovery from a network fault is automatic. By and large, automatic failover can often be implemented to a large extent by network monitoring and reconfiguration scripts. Using automatic failover, recovery times should be measured in seconds.

However, in very complex networks, it may be deemed that automatic recovery is too dangerous. For instance, there may be anticipated cases in which the network is intermittent or is in a "not-sure" state. In these cases, the cause of the problem must be manually analyzed and the proper recovery procedure then implemented.

Manual recovery can take minutes to hours (if appropriate technical or management people are not readily available). Such a delay seriously impacts the ability to achieve extreme reliabilities. Therefore, if manual recovery is to be used, it is imperative that good monitoring and diagnostic tools be in place, that failover procedures be fully documented and practiced, and that the available personnel can be contacted on short notice.

Fault Repair

Once a network has been reconfigured to route around a fault, the fault must be repaired quickly. The longer the fault remains, the more likely it is that a second fault could take down all or part of the network.

Fault repair not only includes equipment repair and testing but also the return of that portion of the network to service. This means notifying all interested components that this portion of the network is now available for use.

235

As we have already described, in some configurations network restoration is automatic. In others, it is a manual process which will further lengthen the time that a backup is not available, thus increasing MTR and therefore reducing availability because of the increased risk of a total network failure.

Transaction, Session, and Connection Loss

When a network failure occurs, there is likely to be activity which is lost and which must be reconstructed. This activity is hierarchical in nature – transactions sent over sessions established over connections.

- *Transactions* comprise a bounded set of business and database activity which is either totally committed to the database or totally rolled back (the atomic characteristic of a transaction). If a transaction is interrupted, it may be aborted and then must be resubmitted.

- Transactions are sent over a *session* between the client and the service. Typically, a session is created when a client logs on to a service. Thereafter, all traffic (typically transactions) flows from the client to the service over this session. If a session is lost, any transactions in progress are aborted. The session must be reestablished by relogging on.

- A session is established over a *connection* (typically a TCP/IP connection) that is created upon the request of a client. The connection is made to a particular service by specifying the IP address of the host and the port number of the service on that host. Once the connection is made, the client can establish a session with that service by logging on to it (unless the service does not care to whom it is speaking, in which case the connection becomes the session). If the connection is lost, the session is lost; and both must be reestablished.

It is important that one understand the loss profile of transactions, sessions, and connections in the event of a network fault. If a transaction is aborted, it must be resubmitted. If a session is lost, the client must log on again and resubmit any uncompleted transactions. If a connection is lost, the connection must be reestablished, the client must relog on and uncompleted transactions must be resubmitted.

By properly structuring the client application and server service, any of these losses may be programmatically recovered without the user being aware. Otherwise, the user must be notified of any of these conditions so that he may take the appropriate action.

Any of the automatic network recovery mechanisms that we have previously described can, in principle, protect transactions, sessions, and connections from loss. However, this depends upon the time that it takes to recover. After a while, various timeouts may trigger such losses.

In any event, such losses are inevitable. There must be procedures, manual or automatic, to recover from them.

Cost

As we have seen, reliable networks can carry a significant cost in terms of redundant equipment, paths, and facilities. Certainly, to the extent that existing networks with alternate routing capability can be used, redundant network cost can be as little as that for redundant interfaces to the network.

In the more general cases, redundant LANs and WANs with all of the requisite routers, bridges, switches, and hubs must be acquired.

Added to this must be the monitoring and diagnostic tools and the cost of network management by the network administrators.

In some cases, it may be deemed that the cost of redundant networks is not worth the added reliability that they bring. Such tradeoffs may result in acceptable single points of failure for certain subnets. Staying with single LANs is a common example of this as LANs are generally deemed inherently reliable (until one gets accidentally cut or unplugged). However, be careful here. Studies have shown[76] that campus LANs "are in a permanent state of disrepair."

Remember, your network can have high performance, low cost, and high reliability. Pick any two. No one ever said that reliability is inexpensive.

A Case Study

An interesting example of the use of redundant LANs is the train control system used by Amtrak along the Northeast Corridor of the United States. A fault-tolerant HP NonStop server provides tracking and routing services to controllers sitting at about two dozen workstations (Figure 5-5). The train controllers are presented with a graphic view of the territory that they are controlling along with its current state. Current state includes such things as train positions, switch settings, lights, hazards, and so forth as communicated from the field.

A key issue in the initial design was responsiveness. A change of state in the field (such as a train movement) had to be reflected in the controller displays within 500 milliseconds. Similarly, an action by

---

[76] *Keeping Systems Running*, Connexion 2; September 1992.
Lan Downtime Clear and Present Danger, Data Communications; March 21, 1990.
Note: These pronouncements are dated, but the warning is still applicable.

the controller (such as throwing a switch) had to be communicated to the field within 800 msec. There simply wasn't time for the workstations to poll the server for state updates and to interact with the server in the complex process of validating and sending a command to the field. Furthermore, the effects of a command had to be displayed on all workstations simultaneously.

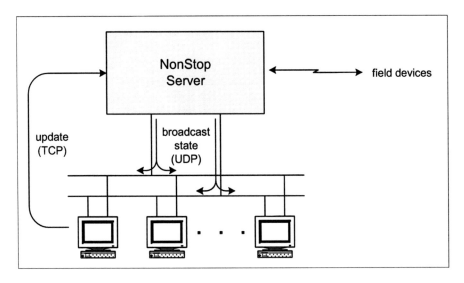

Amtrak Train Control System
Figure 5-5

The solution was to provide every console with the complete state of the railroad so that all console work could be done locally and only the result sent to the server for execution.

As shown in Figure 5-5, the workstations receive state changes from the server via a UDP broadcast (these are datagrams sent to all connected devices). To provide high reliability, the broadcasts are sequence numbered and are sent over a dual rail LAN. Each

239

workstation receives all traffic, sequences the messages, discards duplicate messages, and requests resends via a TCP connection.[77]

When a workstation has a field command to issue, it sends the command over the TCP connection to the server. The server will issue that command to the field and will broadcast any resulting change in state to all the workstations simultaneously, including the workstation that issued the command.

A LAN failure is transparent to the system since workstation/server communications will continue unaffected over the surviving LAN. The health of the LANs and workstations are monitored by heartbeats sent by the server to the workstations over the TCP/IP channel used for commands.[78] If one heartbeat to one workstation fails, that workstation has probably suffered a network interface card failure. If both heartbeats to a workstation fail, the workstation is probably down. If all heartbeats over one LAN fail, the LAN is probably down.

The heartbeats continue even during a LAN failure and are used to detect when a failed component is returned to service.

What's Next

In this chapter, we have discussed how to interconnect reliably the components of an active/active system. Users must be connected to their nodes, nodes must be connected to remote databases, databases must be interconnected for replication, and nodes must be interconnected for heartbeats. If any one of these connections should fail, the active/active network must immediately be reconfigured to

---

[77] See the description of the NetWeave middleware facility at www.netweave.com.

[78] Contrary to what we said earlier, in this case the server can be the initiator of heartbeats since its client base is fixed and known to it.

circumvent this fault. Reconfiguration may entail additional networking to reassign users at a failed node to a backup node.

Of equal importance is the reliability of the databases and the nodes in the active/active network. In the next chapter, we talk about recovering from database failures. Following that, we talk about recovering from node failures.

Chapter 6 - Distributed Databases

"Two heads are better than one."

- conventional wisdom

In Chapter 3, <u>Building Active/Active Systems</u>, we explored the characteristics of distributed databases in some detail. In this chapter, we review the needs for distributed databases and the issues associated with them. We then deal with the impacts of a database failure and the mechanisms to recover from it.

The Need for Distributed Databases

The purpose of an active/active system is to provide extreme availability – six 9s or better. Since we are building these systems from components that might typically have three or four 9s availability, there must be redundancy built into the system so that it can survive any single failure (remember Rule 2 from Volume 1, Chapter 1 – *providing a backup doubles the 9s*[79]).

Furthermore, this redundancy must be distributed across the network so that should one component fail, the system survives. This applies to the processing nodes, the interconnecting network, and the application database.

Thus, at least one additional replicate of the database must be accessible to each of the processing nodes in order to protect it from the loss of access to its primary database. Should a node lose access to all copies of the database, then it is, for all practical purposes, down. Loss of access can be caused by a database failure, a node failure, or a network failure.

---

[79] See Appendix 1, <u>Rules of Availability</u>.

An additional need for distributed databases is locality of data. If a user's transactions need to access data that is not local to his node, then communication delays can add significantly to response time. By having copies of the database at each node, it is guaranteed that data needed by a transaction is always available locally. However, providing database locality to each node can seriously impact the cost of the system.

Finally, there are many cases in which it is desirable to have database copies at other systems for compute-intensive or database-intensive operations. Query processing is a good example of this because complex, long running queries can significantly load a system. In these cases, it is advantageous to keep up-to-date copies of a master database on one or more satellite nodes on which the query or other processing can be run.

Database Synchronization

The replicate copies of the database across the network must be kept in synchronization so that all users see the same application state no matter which database they are accessing. There are several methods for maintaining synchronization, including[80]

- asynchronous replication
- synchronous replication
    - coordinated commits
    - network transactions (dual writes)
    - replicated lock management
- transaction replication

---

[80] See Chapter 3, <u>Building Active/Active Systems</u>.

Asynchronous Replication

Asynchronous replication[81] distributes changes after the fact. That is, once a change is made to a source database, it is sent to the other database copies in the network. Therefore, the view of the application's state may be different at any given point in time depending upon which database is being accessed. However, all views are consistent and catch up to the other views in a short period of time. This time is known as *replication latency*.

Applications are unaware of the replication activity because the replication engine is decoupled from the application. Therefore, application performance is unaffected.

Asynchronous replication carries two problems with it:

- It is subject to data collisions when updates are made to the same row in different nodes at about the same time. In this case, different updates to the same row are replicated. Data collisions should generally be detected and then resolved, either automatically if possible or else manually.[82]

- It is also subject to data loss in the event of a failure. Any changes in the replication pipeline at the time of the failure will be lost until the failed system has been restored and transactions recovered (assuming the transactions are still available after the failover).

---

[81] See Volume 1, Chapter 3, <u>Asynchronous Replication</u>.
[82] See Chapter 4, <u>Active/Active Topologies</u>.

Synchronous Replication

Synchronous replication[83] ensures that a change made to a source database is made to all databases in the network at the same time or is not made at all. The view of the application state is the same regardless of which database is being used.

With synchronous replication, there is no problem with lost data or data collisions. However, synchronous replication is not as transparent to the application as is asynchronous replication. The application must wait until a transaction has completed or is safe-stored on all remote nodes before it can return a completion status to the committing application. This time adds to the application response time of the system and is called *application latency*.

In Chapter 3, <u>An Active/Active Primer</u>, we described three forms of synchronous replication – coordinated commits, network transactions, and replicated lock management:

- With network transactions, a global transaction is started across the network that includes all database copies in its scope.

- With coordinated commits, changes to the source database are replicated to the target database via asynchronous data replication, transparent to the application. However, the transaction is not committed until all database copies have indicated that they are ready to commit.

- With replicated lock management, a lock on each data item is first acquired across the network from all database copies. Updates are then replicated via an asynchronous replication

---

[83] See Volume 1, Chapter 4, <u>Synchronous Replication</u>.

engine. Once all updates are made, the transaction will not be committed until all database copies have indicated that they are ready to commit.

Transaction Replication

Rather than replicating data, one can replicate transactions. With transaction replication, all systems execute all transactions to keep the databases in synchronism. Transactions may be routed to all nodes through the network, or they may be routed to local nodes and forwarded from these nodes to the other nodes in the application network.

Transaction replication has the following characteristics:

- It is not scalable, since every node must execute every transaction. Therefore, it is not truly an active/active configuration. In active/active configurations, every node is processing a different set of transactions.

- There is a possibility of its own form of data collisions. This occurs if the result of processing a transaction is different on different nodes, leaving the database copies in different states. These collisions can be very difficult to detect, and must usually be resolved manually.

Issues with Distributing a Database

There are several issues associated with distributing a database. They include:

- replication latency
- application latency
- data loss following a failure

- deadlocks
- collisions
- database of record
- remote access
- recovery

The first few issues are summarized below. Except for database of record, these issues have been covered in some detail in Volume 1 of this series and have been summarized in Chapter 3 of this volume, Building Active/Active Systems. They are reviewed below and are followed by a more in-depth discussion of remote database access and data recovery.

Replication Latency

Replication latency applies both to asynchronous and synchronous replication. It is the time that it takes to post a change to the target database once it has been made to the source database.

However, the effects are different for asynchronous replication and for synchronous replication. For asynchronous replication, should a failure occur, any data in the replication pipeline will likely be lost. If it takes 500 milliseconds to replicate a change, then up to 500 milliseconds of data may be lost due to a failure. Also, during the interval that it takes for a change to be replicated, there is a chance for a data collision caused by two nodes updating the same row, with both nodes being unaware that the other node is doing the same. The longer the replication latency, the greater is the probability of a data collision.

For synchronous replication, neither data loss nor data collisions are a problem. However, the synchronous replication processes described above all delay the completion of a transaction for one reason or another. This delay represents a performance impact and is called application latency.

Application Latency

We have already hinted above at application performance issues. The question is to what extent is an application affected by distributing it? If an application is slowed by the time it takes to update all copies of the database, this is called *application latency*.

So far as the replication technique is concerned, there is no impact on application performance when using asynchronous replication since the replication engine is decoupled from the application (except for the usually secondary effect of increased loads on the databases, nodes, and networks comprising the application network). Replication is done asynchronously with respect to the application, which is otherwise unaware of this activity.

However, all synchronous replication techniques, by their very nature, impact the application by introducing application latency, as described above. With coordinated commits, the application must wait for the transaction to commit across the network. With network transactions, the application must wait for each update to complete across the network. With replicated lock management, the application must wait for each lock to be acquired across the network and for the transaction to be committed across the network.

It has been shown that network transactions may have better performance if transactions are small, transaction rates are low, and nodes are collocated (that is, the network latency is low).

On the other hand, coordinated commits are advantageous for large transactions, high transaction rates, or if the nodes are geographically dispersed (see also Chapter 10, Performance of Active/Active Systems, in Volume 3).

Data Loss Following a Failure

Replication latency when using asynchronous replication is the major impact on RPO – the recovery point objective. RPO is the stated goal of tolerable data loss for an application.[84]

To minimize RPO, it is important to choose an asynchronous replication engine with a small replication latency or else to use a synchronous replication engine (or another implementation, such as split mirrors, as described in Chapter 16, Related Topics and Drivers, in Volume 3). The replication latency of a data replication engine is typically controlled by the number of disk queuing points within the engine, polling delays, communication buffering, network latency, and transaction synchronization for referential integrity (see Chapter 10, Performance of Active/Active Systems, in Volume 3). Replication latency time generally runs from a few hundred milliseconds to several seconds for asynchronous replication.

Deadlocks

Network deadlocks are not introduced by using asynchronous replication. However, network deadlocks with synchronous replication or network transactions are no different than normal deadlocks in a single database configuration.[85] Basically, two application instances, either in the same node or in remote nodes, are attempting to lock the same set of database items in a different order.

There are several common solutions to deadlocks:

- *Deadlock Protocol:* Make sure all applications are locking data items in the same order. This procedure is called an intelligent locking protocol (ILP); that is, one locks data items

---

[84] See Volume 1, Chapter 6, RPO and RTO.
[85] See Volume 1, Chapter 8, Data Conflict Rates.

in a particular order to avoid deadlocks. Depending on the application, this may need to include a global mutex as defined below.

- *Timeout*: Each deadlocking application releases its locks and retries later at a random time.

- *Global Mutex*: Provide a mutually exclusive object that must be seized before a group of data items can be updated. This mutex must be global to all applications. For instance, all applications might need to lock a particular object in a particular node before updating a related set of data items.

A mutex for a distributed database is a little different than that for a single database. It must reside on one node, and all other nodes must reach across the network to acquire it. Provision must be made to move the mutex should its node go down.

Collisions

Data collisions are a problem only for asynchronous replication. Data collisions are analogous to synchronous deadlocks in that they occur when different application instances on different nodes update the same data item in their respective local databases at the same time. In the case of asynchronous replication, each local update will be successful and will be replicated to the other databases, leaving all database copies in an erroneous and inconsistent state.

Data collisions can be automatically detected; and in some cases they can be automatically resolved, as described in Chapter 4, Active/Active Topologies. Otherwise, they must be manually resolved.

Data collisions can be prevented by using synchronous replication.

Database of Record

An important problem in today's distributed databases and data warehousing technology is termed the "single version of truth." Over a period of years, and due to mergers and acquisitions, an enterprise typically develops a number of operational systems and "data marts" owned by various organizations to support their activities. However, the same data item (such as a name or address) may be different in these databases. What is the true value? The common solution to this problem is to build a common centralized data warehouse that incorporates all of the corporate data and becomes the "single version of truth."

Active/active systems have a similar problem, one which is called the "database of record." In an active/active system, there are multiple copies of the database distributed across the application network, all of which are supposed to be identical. But are they really? If asynchronous replication is being used, then the copies will diverge slightly due to replication latency. They may also diverge due to errors or software faults. Many regulations require that a specific database be identifiable as the single database of record and that its data be considered the "single version of truth."

This leads to the following dilemma in an active/active system. What if the copies of the database in the network have diverged somewhat? Which database is the correct one?

Typically, one specific database in the network is taken as the database of record; and it is declared (rightly or wrongly) to be the correct one. However, should it fail or be inaccessible due to a node or network failure, some surviving database must be designated the database of record.

There are several active/active topologies and algorithms that can aid in the solution to the following dual aspects of this dilemma:

- Which database copy should be chosen as the database of record?

- Which database should take over that role should the primary database of record become inaccessible?

Synchronous Replication

If synchronous replication is being used, all databases are guaranteed to be in the same state (short of a fault of some sort); and any one can be chosen to be the database of record. In the event of the need to promote another database to the database of record, any of the surviving databases may be used.

In a sense, synchronous replication is akin to the use of mirrored disks in monolithic systems. Should one disk go down, its mirror is trusted to be an exact replica.

Partitioned Database

There are many active/active architectures that can take advantage of database partitioning (see Chapter 4, Active/Active and Related Topologies). Partitions are also called *shards*. In such cases, though there are multiple copies of the database in the application network, updates for any particular row are always directed to the same database copy – that which contains the designated partition holding that row. In this way, data collisions are avoided. The updates are then replicated to the partition copies.

In this case, the database of record is itself partitioned across the database copies, with each primary partition serving as the database of record for its data. Should the database containing a primary partition fail, another partition must be selected to act as the primary partition.

253

Master/Slave Topology

In a master/slave (or hierarchical) configuration, all updates are made to a designated master copy of the database in order to avoid data collisions. The updates are then replicated to the slave database copies. In effect, the master database is the primary "partition" for the entire database and is the database of record.

Should the master database fail, one of the slave databases must be promoted to become the new master.

Global Mutex

If a global intelligent locking protocol (ILP)[86] is used, global mutexes (such as global locks) will be resident in one database copy. Any node which wishes to update a row must first obtain the global mutex protecting that row. It may then proceed to update the data in its copy, and these updates are distributed to the other copies (including the one holding the global mutexes) via data replication. In this way, it is guaranteed that there will be no collisions.

The database holding the global mutexes should be chosen as the database of record. Should that database become inaccessible, the global mutexes must be moved to another database.

General Asynchronous Replication

In the general active/active asynchronous replication case in which all nodes are equal, there is no reason to choose any particular database as the database of record. However, once chosen, the node containing the database of record should be the node that will win all data collisions.

---

[86] See Volume 1, Chapter 4, Synchronous Replication.

If that node fails, another node must be chosen as the database of record.

Verification and Validation

Verification and Validation utilities are available to ensure that two database copies are in synchronization and to bring them into synchronization if they are not. Such a utility should be run periodically in order to compare each database copy to the database of record. If there are differences, the database copy must be corrected so that it matches the database of record.[87]

Succession

Should a database of record become inaccessible, it is necessary to promote another database copy to become the database of record. If there are only two database copies in the network, then the choice is trivial – the surviving database becomes the database of record.

Otherwise, if synchronous replication is being used, any database in the application network may be promoted. If there are multiple surviving copies, the decision might be based on location, facilities support, or other criteria.

If asynchronous replication is being used, there is always a window of uncertainty (the replication latency) during which some copies of the database may have received the latest updates and others may have not. If it can be deduced which database copy is likely to be the most up-to-date, that copy should be promoted.

Alternatively, if a Verification and Validation utility is being used, the most recent database copy to have been synchronized with the database of record could be chosen as the new database of record.

---

[87] See Chapter 13, SOLV.

Issues with Remote Access

Why Remote Access?

Having to access a database which is geographically remote brings with it its own set of issues. The need to remotely access a database may be a consequence of several factors. For instance, as shown in Figure 6-1:

- *Topology:* In a multinode network, there only need to be two copies of the database to achieve the desired level of availability since mirrored copies are the norm in highly-available systems. Therefore, all but two nodes may have to access needed data over the network. In Figure 6-1a, Node 1 must gain access to a database copy via Node 2. Node 3 gains access via Node 4.

- *Network Attached Storage:* Another topological issue is network attached storage (Figure 6-1b). The application's database may be kept on redundant network attached storage (NAS) devices. These are storage systems that are not directly connected to any particular node but rather are connected to the network. They are accessible to any node in the network.

  Presumably, each node will use the NAS device which is geographically closest to it, except in the event of a database or network failure. The NAS devices may be members of a SAN (storage area network), in which case they are shared by other systems as well. However, this does not affect their use in active/active configurations except for performance.

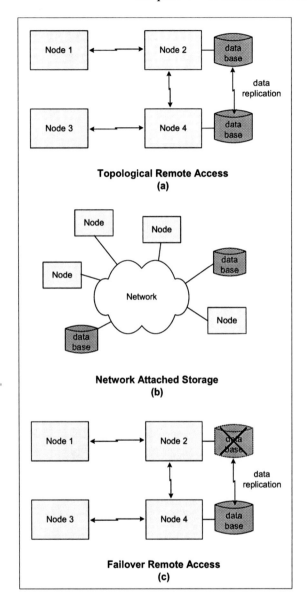

**Reasons for Remote Database Access
Figure 6-1**

- *Failover*: In the event of a database, node, or network failure, access to the database normally used by a node may be lost; and another database must be used. In Figure 6-1c, the database connected to Node 2 has been lost. Node 2 must now reach out to Node 4 for database access. Even worse, Node 1 must route its requests through Node 2 to Node 4.

Issues that must be dealt with when faced with remote access of data include:

- network latency
- data locality
- partitioning
- network redundancy
- security
- fairness

Network Latency

There is an additional performance issue independent of the method of synchronization, and that is network latency. If a node must use a remote database (either normally or as a result of a failure), each database command must propagate over the network, as must its response. Round-trip time for a signal is about 2 milliseconds per 100 miles (roughly half the speed of light over fiber or copper). Thus, if the database is 1,000 miles away, 20 msec. will be added to each database action – even if it is a fast 5-microsecond read from disk cache.

Therefore, it is important that the network topology satisfy the performance needs of the application, at least in normal operation. Database distances of a hundred miles or so are probably acceptable since transaction times may be increased by only a few milliseconds. However, database distances in excess of a thousand miles will probably lead to unacceptable response times as the better part of a second may be added to a transaction. In this case, it may be necessary to add database copies in order to increase data access locality so that the desired performance may be obtained. This, of course, will increase overall system cost. It is those distances that lie between a hundred miles and a thousand miles that must be analyzed carefully with respect to the need for database locality in light of the performance/cost tradeoff.

Data Locality

If a transaction must reach across the network to access data, network latency will increase the transaction response time of the application. Therefore, maximum performance is obtained when each node has a local copy of all of the data it needs. This argues for a copy of the database at each node.

Of course, this can be a costly configuration. The availability goals of active/active systems are achieved if there are at least two copies of the database in the application network. Providing more copies to satisfy data locality is a cost/performance tradeoff.

Partitioning

There are some applications in which an efficient compromise can be reached between data locality and cost. In many distributed applications, users at any one site may predominantly access only a portion of a database; and these portions do not overlap across the various nodes. For example, a sales application may support regional sales offices. Sales people in a region will, for the most part, access and update only that data associated with their region. They will only occasionally have need for data in another region.

In this case, each node hosts that part of the data that is associated with its local users, as shown in Figure 6-2a. Then for most of their work, users' transactions will be handled by the local database. They will usually read locally but will have to write to all databases across the network.

To provide redundancy, each local database partition is backed up by another (remote) node. Then only if a local database should fail will these local database accesses flow across the network. In this configuration, even though every node has a database, there are only two copies of the total database across the network. That is, rather than requiring a full database at each node, only the local partition plus the backup for one remote partition is required, thus reducing the network database requirements and cost significantly.

Likewise, should a node fail or become isolated by network faults, the users at that node can be switched to the node which is backing up the database portion related to those users and can continue to have locality of access (Figure 6-2b). In this case, performance is perhaps better than that achieved when accessing data remotely since only the transactions need flow over the network, not each database access.

The configurations described above are forms of database partitioning. Data is partitioned by use and is located where it is most needed.

In certain circumstances, partitioning can provide even greater benefits. If the application is such that only local users can update their partition, and if all remote users are limited to read access only of that partition, there will be no data collisions. In unpartitioned busy systems with significant replication latency, collisions can happen frequently and, if manual resolution is required, can add significant operating cost to a system.

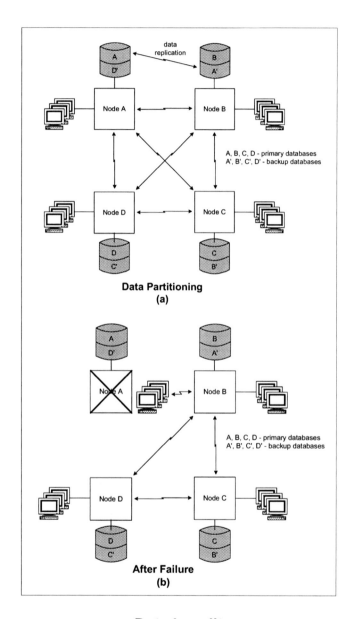

Data Partitioning
(a)

A, B, C, D - primary databases
A', B', C', D' - backup databases

After Failure
(b)

Data Locality
Figure 6-2

The restriction that limits updates to only local users can be relaxed by requiring that remote users update only the primary partition. That is, they cannot update a backup copy that may be closer to them and let data replication propagate the changes to the primary partition. All users must update the same partition copy, no matter where it is. In this case, data collisions will be avoided.

Assuming that the need for remote access is infrequent, database partitioning brings the no-collision advantage of synchronous replication to an asynchronous replication environment while substantially preserving the performance advantage of asynchronous replication.

Network Redundancy

In order to achieve the extreme availabilities which we are seeking, not only must the databases be redundant, but so must the networks interconnecting them. So far as databases are concerned, this is primarily the replication link joining each pair of databases.

Chapter 5, <u>Redundant Reliable Networks</u>, explored redundant networks and their configurations, failure modes, and recovery in some detail. Given redundant links between nodes for the purpose of replication, these same links can be used for remote database access in the event of a database failure and for user traffic in the event of a node failure, assuming sufficient communication capacity is available. Note that these uses are mutually exclusive – the network will be used for only one of these at any given time.

Security

It is one thing to own all of the networks used by a system because they are local. It is another thing to use public network facilities. In this case, one must be concerned about compromising sensitive data either by snooping, spoofing, sniffing, or modification. Snooping is the monitoring of data by an outsider. Spoofing is an outsider entering erroneous data as if he were a valid user. Sniffing is the monitoring of packet data by a network analyzer. Modification is altering data that is otherwise flowing legitimately over the network.

The common protection of data that must be secured over public (and sometimes even private) networks is data encryption. Data encryption should be seriously considered whenever public networks are used to interconnect nodes in an active/active system.

Another problem is denial-of-service attacks. The best defense against such attacks is a good firewall isolating the system from the public network.

Fairness

In some applications, it is imperative that all users are treated equally. This has a particular impact when it comes to response time of networked transactions. Users with local data access will have much faster access to the database than users that must reach that data across the network.

A good example of the fairness requirement is a trading application. Traders in London, New York, and Hong Kong must be given the same access to the database of a New York-based stock exchange in order that they can all trade on an equal footing. Architectures for fairness are discussed in Chapter 4, <u>Active/Active Topologies</u>.

Failure Modes

Access to a database may be lost for several reasons:

- The database itself may fail.
- Its host node may fail.
- The network providing user access to it may fail.

If any of these events occur, the affected users must be provided with access to an alternate database in order to continue service to them.

Failover

Redundant Configurations

To recover from these faults, redundancy is required at several levels, as shown in Figure 6-3.

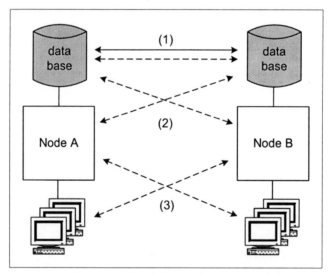

Redundant Database Access
Figure 6-3

(1) *Data Replication Channel:* The first requirement for redundancy is the data replication channel. If that channel is

lost, database isolation occurs (the so-called split-brain syndrome).

(2) *Alternate Database Access:* If a node loses its database, it can continue in operation by accessing a remote copy of the database.

(3) *User Switching:* If a node loses its database, or if the network latency to a remote database for each database operation is unacceptable, the users can be switched so that they have direct access to a node with an operational database. Users can also be switched to a surviving node in the event of a node failure.

Database Failure

If a database copy should fail, all users of that database must be given access to a surviving database. One way to do this is to remap the configuration of those applications that were using the failed database so that they are now pointing to another copy of the database (Figure 6-4a). This may require taking down the applications, remapping them to the backup database, perhaps recompiling the SQL statements, and then returning the applications to service.

This procedure, of course, requires that there be a path from each node to a backup database (see link (2) in Figure 6-3). This link may take one of several forms.

IP Link

If the database subsystem is an independent system that is reached via its own IP address, even from its local node (Figure 6-4b), then the node with the failed database may replace the IP address of the failed database with the IP address of the backup database. After remapping to that database, applications on the affected node can then continue by using the backup database over the network.

Oracle's SQL*Net is an example of this capability. Remote clients can establish a TCP/IP connection with a listener resident on the database host. These clients can then access the database as if it were local.

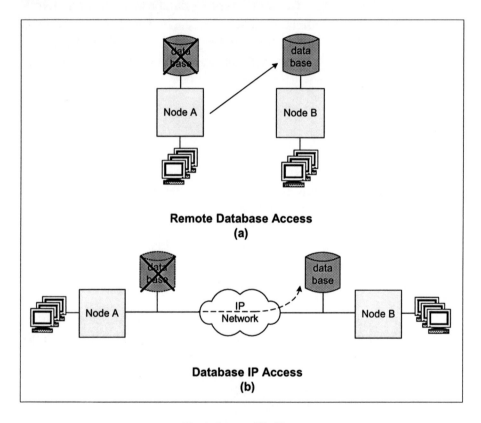

Database Failover
Figure 6-4

Note that IP rerouting is also applicable to network attached storage (Figure 6-5a). With NAS, all databases are remote and are accessed over the IP network. It is just that some NAS devices may be closer to a node than others. A node will generally use the NAS copy closest to it and will switch over to the other NAS copy only in the event of a database failure.

In some cases, it may not be feasible to connect directly to the node containing the backup database. For instance, the network topology might not provide such a path; and traffic may have to be routed through one or more nodes (as shown in Figure 6-1c). If this is IP traffic, intermediate nodes will have to be reconfigured as routers. The requirement for intermediate nodes will impact performance to some extent, though IP routing is an efficient process.

NonStop Expand

A powerful approach to remote database access is the Expand networking capability provided by HP's NonStop servers (Figure 6-5b). Expand is a redundant network that provides a single system image across all nodes of a distributed system.

With Expand, an application running in one node has access to all of the resources across the network just as if those resources were local. Thus, in the event of a database failure, an application need only remap its database access to the backup database and continue on. So far as the application is concerned, the remote backup database which it is now using appears to be local (except, of course, for network latency).

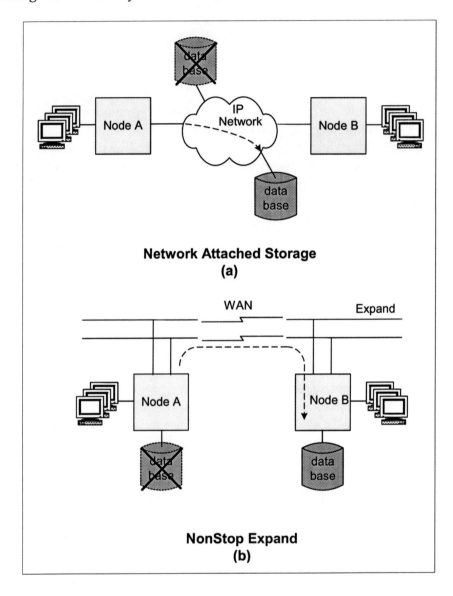

Network Attached Storage
(a)

NonStop Expand
(b)

Database Failover
Figure 6-5

Switching Users

If network latency should be too large to support the application in the event that individual database commands are being executed over the network, then the users can instead be switched over to a remote node with local access to the database. Now only the transaction request and reply will be subject to network latency, not each individual database operation.

This is the same strategy as that for recovering from a node failure, as discussed next.

Node Failure

Should a node fail, the only recourse is to switch all affected users to another node. Typically, users are moved to the closest node to minimize network latency. However, the users may be partitioned so that different sets of users are moved to different nodes for load balancing.

Recovery from node failures is covered in more detail in Chapter 7, Node Failures.

Network Failure

If the communication link over which replicated data flows should fail, then the replicator must switch to its backup link. Communication link redundancy and failover have been described in some detail in Chapter 5, Redundant Reliable Networks.

Shared Links

We have described the need for several communication links required to support database synchronization and to provide backup channels:

- a data replication link
- a link for a node to access a remote database
- a link for users to access a remote node

These need not be separate links. They can all share a common redundant link, as shown in Figure 6-6.

Usually, this link would only be used for data replication. However, should the node need to access a remote database or the users a remote node, they can use this link as well.

In fact, these uses are mutually exclusive. If the link is being used for one purpose, the other uses are dormant. If a database fails, there is no longer synchronization with other databases; and either the node will use the channel to access a remote database or the users will use the channel to access a remote node. The channel should be sized to handle that case which will generate the largest load.

Database Recovery

Failing over from a failed database is only half the story. A repaired database must also be returned to service. Prior to doing so, it must be resynchronized with the current application database. The resynchronization procedure depends upon how long the database has been down.

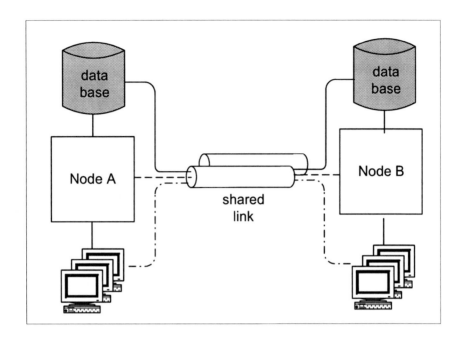

Shared Links
Figure 6-6

Replication Queue

If the database has not been down very long (how long is "long" is addressed in the next section), then the database can be brought up-to-date by draining the replication queue.

While the database was down, systems that needed to replicate changes to it have been queuing those changes. When the database comes online, the queued changes are released and flow over the replication channel to update the recovered database.

As the queue of stalled changes is drained, new changes, of course, are being added to it. When the queue of changes has settled to a length commensurate with normal operation, the database can be considered to be recovered. At this point, the applications running on the now recovered node can be switched back to their local database; or the users can be switched back to their home node, whatever the case may be.

Reload

An alternate technique is to simply start fresh and copy the active database to the repaired database. We asked above "how long is 'long'?" The general answer is this: if it is going to take longer to drain the queue than it is to simply copy over the entire database, then the database should be recreated by copying.

The copy must be an online copy that does not require pausing the applications, or else users will have no service for what could be a significant time. In addition, the copy ideally will continue to post changes to the recovering database so that there is not a large queue of changes to post following the copy. Posting changes could take as long as the copy.

Online copy utilities with these characteristics are described in Chapter 8, Eliminating Planned Outages with Zero Downtime Migrations (ZDM), and in Chapter 12 of Volume 3, SOLV.

What's Next

We have seen a situation in this chapter in which a node might be abandoned – that is, its database has failed. But a node will certainly be abandoned if it should fail.

In the next chapter, we discuss node failures – how nodes can fail, how to detect a failed node, how to fail over from a failed node, and how to restore a node to service.

Chapter 7 - Node Failures

"Success consists of going from failure to failure without loss of enthusiasm."
- Winston Churchill

Causes of Node Failures

There are many events that can take a node out of service (see Figure 7-1):

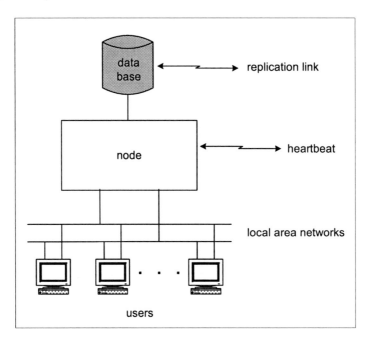

Node Environment
Figure 7-1

- The *local network* connecting the consoles to their node may fail.
- The *replication link* to remote nodes may fail.
- The *database* may fail.
- The *heartbeat network* used to monitor nodes may fail.
- The *node* itself may fail.

Each of these conditions must be detected, and a failover strategy (automatic or manual) must be executed.

Detecting Failures

Every component that can cause a node to fail must be monitored so that failures can be immediately detected. In the event of a failure, the component must be removed from service and replaced with an operable component. Finally, the component must be repaired and returned to service.

Much of this topic has been covered in Chapter 5, <u>Redundant Reliable Networks</u>, and Chapter 6, <u>Distributed Databases</u>. We will refer often to those chapters and fill in the holes.

Local Network

Should the network connecting the user workstations to the node fail, the users can no longer pass transactions to the node. Local networks are typically implemented as a dual rail LAN. The health of a local network rail can be detected by:

- heartbeats sent between the node and the user workstations.

- the lack of a response to a client query.

- monitoring of network statistics.

These fault-detection methods are discussed thoroughly in Chapter 5.

Replication Link

Should the replication link fail, the database becomes isolated and cannot be synchronized with the rest of the system. This is the so-called split-brain syndrome, discussed later under "Other Issues."

Replication links are typically long-haul dedicated links (that is, the nodes are geographically dispersed for disaster tolerance or data locality), though they need not be symmetric. For instance, the primary link could be an ATM channel; and the backup could be the Internet.

The fault detection mechanisms for these links is much the same as for local networks – heartbeat, lack of response, and monitoring - and are further described in Chapter 5.

Database

If the database should fail, transactions can no longer be executed locally. A database failure is detected by error responses returned to application requests by the database subsystem. Recovery from database failures is discussed in Chapter 6, <u>Distributed Databases</u>.

Heartbeat Network

A node cannot be self-monitoring as it cannot report its demise if it is dead. Furthermore, if it is sick, it cannot be relied upon to report its ill health. Therefore, it must be monitored by some external entity.

Heartbeats

The only reliable means to monitor a node is for it to periodically report its health or alternatively to be queried periodically regarding its health. These messages are called heartbeats or "keep-alive" messages. If a node has detected a fault, it can report the nature of the fault (the fault might not be fatal – losing one of its network rails, for instance). If it doesn't report or respond, it is probably out of service. In many cases, a simple IP ping is suitable for a heartbeat.

The most effective heartbeat is an end-to-end heartbeat that tests the application as well as the node and network. In this regard, actual data messages are often used to satisfy the heartbeat requirement. Heartbeats are sent only during periods of inactivity.

Note that the loss of a heartbeat or heartbeat response may not be caused by a node failure but rather by a network failure. If it is determined that the network is down rather than the node, it may be desirable to take the node out of service anyway to avoid the split-brain syndrome described later.

Heartbeat Period

In many systems, missing one heartbeat or heartbeat response is not a great cause for concern.[88] The heartbeat may have been lost by the network, or the heartbeat may have been delayed due to the node being too busy (neither is a good situation and should be corrected). For instance, the implementation might require the loss of three consecutive heartbeats before declaring the node in failure.

There is a tradeoff between the heartbeat period and heartbeat channel capacity. If a heartbeat is sent every three seconds, and if

---

[88] If heartbeats are sent via TCP, one missing heartbeat will kill the link since TCP guarantees delivery.

three heartbeats must be missed, then it will take up to nine seconds to detect a node failure. The faster the heartbeat, the faster is node failure detection. However, what also results is a greater load on the network and the participating nodes and a greater impact of transient conditions such as network congestion.

However, unlike user-workstation heartbeats, we are dealing typically with only a handful of nodes rather than hundreds of workstations. Therefore, network loading is not an issue; and frequent heartbeats are in order.

Network Redundancy

The heartbeat network is probably the most critical part of an active/active network. The loss or unreliable functioning of the heartbeat network can be catastrophic. Not only can node failures no longer be detected, but two nodes can each decide that the other is down and try to seize the other node's users if automatic failover is implemented. The resulting tug of war might possibly make participating nodes nonfunctional.

Therefore, the heartbeat network should be a highly reliable, redundant network. Moreover, if two heartbeat networks are used, it is prudent to send heartbeats over both networks to ensure they get through.

The heartbeat network is one case in which the Internet might be a viable channel, either for the redundant network or for the backup channel.[89] The delay time is not as critical as it is for transactions or data replication, and security issues are not as severe (though certainly existent – you don't want a hacker to spoof a node failure).

---

[89] See Chapter 5, Redundant Reliable Networks.

If the Internet is used for the backup channel, then the heartbeat period over that channel might be different than it is for the primary channel. For instance, if the primary channel is operable, the heartbeat interval might be one second, thereby leading to a three-second failure detection time if three heartbeat losses are required. The Internet heartbeat interval might be ten seconds. Thus, should the primary channel fail, it could take up to thirty seconds to detect a node failure. This condition of extended failure detection time only exists during the primary channel outage and, while annoying, is probably not catastrophic in most cases because of the small probability of a node failure.

If the Internet is used, the weak link is probably not the Internet backbone but rather the local ISP.

The concept of slower heartbeats is useful for other types of heartbeat backup channels which trade capacity for cost. Dial-up lines and lower speed dedicated links are examples of these.

Failure Detection

Detecting a failure in a heartbeat link is a little different from detecting a failure in some other network link. Typically, a link's health is monitored by sending heartbeats over it during idle times. However, that is the normal activity of a heartbeat link. If the heartbeat fails, it is assumed that the receiving end is down.

This is the importance of sending heartbeats over both rails of the heartbeat link. If heartbeats are being received on one rail but not on the other, then the silent rail can be assumed to be down.

The heartbeat sender can declare a rail failure if its heartbeat messages are rejected or not responded to. Both ends can detect network problems by monitoring the network's performance statistics.

Node

Heartbeats are the key to detecting node failures. Basically, if a node stops sending or responding to heartbeats, it is declared down.

We consider two heartbeat configurations – self-monitoring and arbitration.

Self-Monitoring

In a self-monitoring configuration, the nodes themselves generate and listen for heartbeats. There are two ways that this can be done – by multicasting heartbeats to all nodes or by sending heartbeats only to a backup node.

Multicast:

In a multicast configuration (Figure 7-2a), every node multicasts a heartbeat periodically (such as via UDP broadcast datagrams) in order to announce that it is alive; and every node monitors the heartbeats of the others. Every node is a member of the heartbeat multicast group.

If a node misses the prescribed number of heartbeats from another node, then it concludes that the node is down. It takes the appropriate action, such as switching users from the failed node to itself, if it is the backup node.

The danger in this arrangement is the tug-of-war syndrome mentioned earlier should two nodes which are backing up each other decide that the other has failed. Realizing that this will usually be caused by a heartbeat network failure, there are some protections that one can take to avoid a tug of war:

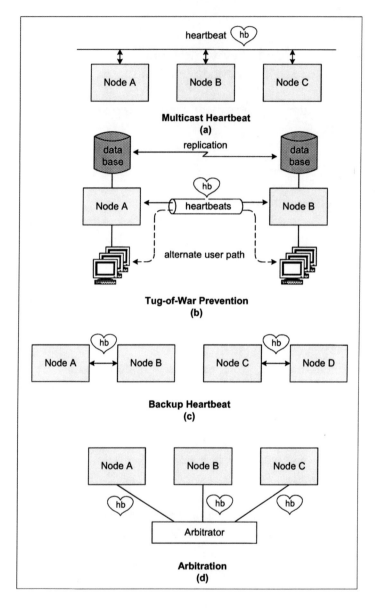

Node Monitoring
Figure 7-2

- If there are multiple channels between the nodes for other purposes (such as heartbeat, data replication, or user switching), queries can be made over these channels to ensure that the other node is really down.

- The decision to fail over can be made manually rather than automatically, thus precluding the downed node from trying to regain control.

- If a common channel is being used for all internode functions such as heartbeats and remote user access, then a tug of war cannot happen. The loss of the heartbeat channel also means that the channel used to switch user traffic over to the backup is also down. Therefore, users cannot be seized by the backup node; and the nodes continue to operate independently.

The last point leads to an interesting configuration to avoid tug of wars while maintaining database synchronization. If a common channel is used for heartbeats and backup user traffic, as shown in Figure 7-2b, but a separate channel is used for replication, then in the event of a heartbeat channel failure, neither node can seize the users assigned to the other node. However, replication continues. There is no change in operation except for the lack of both monitoring and failover in the event (presumably unlikely) of a node failure.

Backup:

With multicast heartbeats, all nodes monitor all other nodes. But this may be excessive since typically a node may back up only certain other nodes. It is only necessary for a node to send heartbeats to those nodes that are backing it up. For instance, nodes may be organized in pairs. Should one member of a pair fail, its users are switched to the other node.

In this case, the heartbeat network is simplified. It will no longer need to be a multicast network. Rather, only backup groups need to exchange heartbeats, as shown in Figure 7-2c. In this diagram, nodes A and B are backing each other up, as are nodes C and D.

The anti-tug-of-war configuration described above holds for this configuration as well.

Arbitration

There is another way to avoid tug of wars and other misdiagnosed conditions, and that is through the use of an arbitrator, as shown in Figure 7-2d. The arbitrator is an independent node that is not part of the application network. It may be a dedicated node or an application running on another system.

The arbitrator periodically polls the nodes in the active/active system for their health. Alternatively, the active/active nodes can generate their heartbeats and direct them to the arbitrator. In any event, the arbitrator makes health decisions concerning each node. If it determines that a node should be taken out of service, it will direct that node to go down (assuming that the node can still hear the arbitrator) and will switch that node's users to their backup nodes. It can do this either by sending routing commands to the involved routers or by requesting the backup node to switch the users, as described later. Alternatively, it may issue a directive to the operations staff to manually switch users or to inform the users directly to reconnect.

One advantage of the arbitrator is that it is the repository of system health, especially if each node details its health problems in its heartbeat. It can, therefore, provide a great deal of information to a system management tool or can, in fact, be implemented as a function provided by a system management tool.

Failover

Should a node be taken down by any of the faults described above, then failover to a spare component is required in order to continue service to users.

Local Network

If the local network connecting the users to the node fails, then the backup local network must be put into service, as described in Chapter 5, <u>Redundant Reliable Networks</u>.

In some cases such as parallel use, network failure recovery is seamless. In other cases, automatic or manual reconfiguration of the network is required.

Replication Link

If the replication link fails, its backup must be put into service, as described in Chapter 5. If this link pair is formed from dedicated services, then the sending node must switch to its alternate interface. Presumably, the receiving node is listening on both interfaces.

If a large public or private network with alternate routing is being used, then a path failure is transparent to the node because the network will reconfigure and provide an alternate route through the network. Of course, if the node's interface to this network has failed, the node must switch to its alternate interface.

If there is a total failure of the replication link, then a choice has to be made:

- Continue in operation with the node. This is the so-called "split-brain" syndrome in that the node's database is not being kept in synchronism with the rest of the application database. The split-brain syndrome is discussed later.

- Switch the users to their backup node.

In those cases in which there are two or more databases in the network, it may be possible to reroute the node's replication traffic to another database copy. Changes made to that database will be propagated to the other databases in the network. This procedure can be very complex but can provide the function of redundant replication networks. If such connectivity exists, the replication networks themselves might not have to be redundant.

Database

Should the local database fail, the node can switch to a remote database copy, as described in Chapter 6, <u>Distributed Databases</u>. Alternatively, the users can be switched to another active node. Methods for switching users are described in more detail later.

Heartbeat Network

If heartbeats are sent over both rails of a redundant heartbeat network, there is no failover action required.

In the unlikely case that the redundant heartbeat network is lost, normal operation can still continue. The penalty is simply that nodes will not be monitored. Should a node fail, it must be detected manually and users switched to their backup node manually.

Node

If a node fails, there is no recourse but to switch its users to a backup node. User switchover is discussed next.

Switching Users

We have discussed two reasons to switch users to an alternate active node:

- A node fails.

- A database fails, and users are switched for performance reasons (network latency has less of a performance impact due to transaction delays than it does due to database action delays).

Typically, groups of users will be assigned another active node as a backup node. For instance, for locality purposes, two nodes may back each other up; and in the event of a node failure, the users at that node will be switched en masse to their backup node. Alternatively, subsets of users at a node may be assigned to different backup nodes for load sharing purposes.

We next discuss network topologies that allow users to be switched between nodes. We then explore methods for doing so.

A Network Topology

Intranode Network Topology

A typical configuration for providing the capability to switch users from one node to another has been examined in Chapter 5, Redundant Reliable Networks. There we discussed a dual LAN connecting each group of users to their respective node via a pair of routers. In Figure 7-3, we repeat this configuration for two nodes backing up each other.

Considering for the moment only Node A, we show each console connected to two LANs (Subnet 1a and Subnet 2a). A console might

285

be using only one LAN at a time, it might be load sharing by alternating the use of the LANs, or it might be sending traffic over both LANs simultaneously. If a console finds that it cannot communicate with its node over one LAN, it can switch to the other.

Likewise, a node is connected to two LANs (Subnet 3a and Subnet 4a) via a pair of independent interfaces (shown with IP addresses IP1a and IP2a). A service (e.g., a process) running in the node listens to traffic being received on both of its interfaces.

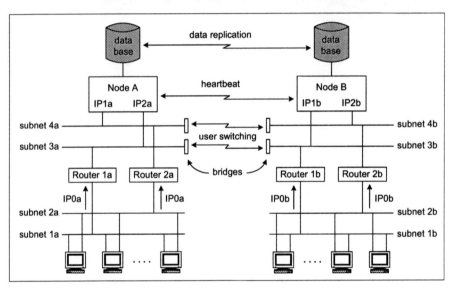

Switching Users
Figure 7-3

The console LANs are connected to the node LANs via a pair of routers. Figure 7-3 shows console Subnet 1a connected to node Subnet 3a via Router 1a and console Subnet 2a being connected to node Subnet 4a by Router 2a.

Virtual IP

The consoles know their node only as a virtual IP address, shown as IP0a in Figure 7-3 for Node A. It is the job of each router to map this to the proper node interface. If a console generates IP0a traffic on Subnet 1a, Router 1a will normally direct that traffic to Node A's interface that is mapped to IP1a. Console traffic on Subnet 2a will be routed by Router 2a to the interface mapped to IP2a.[90]

Thus, a console uses only one well-known IP address to talk to its node and has no knowledge of which LAN is actually being used.

It is this common IP address known to the consoles that is the basis for switching users, as we shall see.

Network Topology

Figure 7-3 shows two nodes, Node A and Node B, each with the intranode topology described above. Node A's Subnets 3a and 4a are simply bridged together with Node B's Subnets 3b and 4b to form a single pair of subnets, Subnet 3 and Subnet 4. A bridge allows only traffic destined for one of its remote subnets to be sent to that subnet. It blocks local traffic so as not to load down subnets with meaningless traffic.

The long-haul links that are now part of these subnets comprise the user-switching network that we discussed earlier. As described then, these links can either be independent links or can share common links with the data replication and/or heartbeat networks.

---

[90] A router actually maps the destination IP address to the interface address of the next hop on its subnet. For an Ethernet LAN, this is the 48-bit MAC address of a network interface card (NIC). A router can map multiple IP addresses to a NIC. A NIC can receive multiple IP addresses. The IP datagram contains the IP address of the ultimate destination.

287

Rerouting Virtual IP

With the internode topology described above, when a failure occurs, all that is needed is to instruct the routers to redirect virtual IP traffic. For instance, Router 1a normally sends IP0a traffic to IP1a on Node A. If it is told to modify its routing tables to send IP0a traffic to Node B's IP1b interface, now all traffic over Subnet 1a will be sent to Node B. A similar action taken on Router 2a will ensure that all of Node A's user activity is now handled by Node B. Users have been switched from Node A to Node B.

Manual Switchover

Users can be switched manually by issuing reroute commands to the routers. This might typically be done from the network management console. Alternatively, the users may be informed to switch over via a command from their console.

Though this will take longer than an automatic switchover, it may be appropriate for complex configurations.

Arbitration

If node health is being monitored by an arbitration system, then it will determine when users should be switched. It will issue the redirection commands to the routers directly. Alternatively, it can issue a directive to the operations staff to switch users.

Automatic Switchover

In IP networks, a node can seize a virtual IP address from another node, thus routing all of that user traffic to itself. The action is simple, but it requires an understanding of IP's Address Resolution Protocol (ARP).

Address Resolution Protocol (ARP)

In order to send an IP datagram to the next hop (which could be an intermediate router or the destination host), IP must map the IP address of the datagram to a device interface address on a subnet to which it is connected. This interface address on a LAN is the hardware address of that device's NIC (Network Interface Card).

To accomplish this mapping, each host and each router maintain a cache that maps IP addresses to interface addresses for the next hop on each subnet to which they are connected. The cache is filled in several ways:

- Every IP datagram contains a sending IP address and a sending interface address as well as the same pair of addresses for the recipient (the interface address is that of the next hop). Since the host or router can listen to all of these datagrams on the LAN, it updates its cache with these sending and receiving address pairs as it listens to each datagram being sent on each of the subnets to which it is connected. Thus, routing tables are self-discovering and automatically maintained.

- If the destination IP address is not found in the sender's (host's or router's) cache, the host or router sends an ARP request over the LAN asking for the interface address servicing that IP address. The appropriate NIC will respond, and that address pair updates the cache in every router on the LAN.

When the IP/interface address mapping is known, the datagram can be sent to its next hop on the LAN.

Gratuitous ARP

A gratuitous ARP is a form of ARP request in which the sender asks for its own address. Depending upon the NIC, the sender may or may not reply (not all NICs can hear their own transmissions). The ARP request will carry the sender's IP address and interface address. If there is an ARP reply (which, of course, can only come from the requester), the reply will carry this address pair as both a sending and a receiving address.

This may seem useless at first, but let us look at its consequences. As we have mentioned, an ARP request contains the IP address and corresponding interface address of both the requester and that system which responded. Every router updates its mapping cache with both the sender and receiver mappings.

Therefore, sending a gratuitous ARP will cause all routers and hosts on the subnet to map the specified IP address to the sender's interface. The sender has seized that IP address.

Implementing the Switchover

The method for one node to seize the users assigned to another node is now obvious from Figure 7-3. If node B wants to take over the users on Node A, it simply sends a gratuitous ARP for IP0a over both Subnets 3 and 4. Since these respective subnets for Nodes A and B are bridged together, the ARP request will flow across the entire subnet; and all routers will change their mapping of IP0a to Node B's IP1b or IP2b, depending upon the subnet.

The result is that now all traffic generated by users at Node A on IP0a is routed to Node B for processing. IP0b traffic generated by

Node B's users is unaffected. Node B is now handling traffic from its own users as well as those originally assigned to Node A.

The topology shown in Figure 7-3 can be extended to the partitioning of users so that some are backed up by one node and some by another node. To do this, each user partition is assigned a unique virtual IP; and the Subnets 3 and 4 are extended to the other involved nodes. Then any node can seize any set of users by issuing a gratuitous ARP for the virtual IP used by those users.

Node Recovery

Should a node fail, its database will rapidly become out-of-date. Therefore, when a node is to be put back into service, its database must first be synchronized with the currently active database for the application. Database recovery is discussed in detail in Chapter 6, Distributed Databases. It may be accomplished either by draining the replication queues which have built up during the outage or by doing an online copy of the active database to the failed database, whichever is deemed to be faster. Online copying is discussed further in Chapter 8, Eliminating Planned Outages with Zero Downtime Migrations (ZDM).

Once the database has been synchronized, all that is required is to switch the users back to their primary node, as described above, in order to rebalance the user load across all nodes.

Other Issues

Several other miscellaneous issues relating to node failures should be discussed.

Manual Versus Automatic Failover

As discussed for other types of failovers, recovering from a node failure can be a manual or an automated process. An automated failover will be much faster, but manual failover may be more practical for complex networks.

A compromise worth considering in some configurations is to failover automatically with follow-up manual tuning.

Load Shedding

If part or all of a failed node's processing responsibilities are taken over by a surviving node, the load on that node will significantly increase. If it is anticipated to increase to the point that response time may be seriously impacted, one should give consideration to load shedding. That is, noncritical processing is stopped until the failed node is returned to service and has taken over its normal processing functions.

Split Brain

If a node with a local database copy should lose its replication channels, its database can no longer remain in synchronism with the rest of the application network. It is now in a mode referred to as "split brain."

There are two ways that the split-brain syndrome can be handled.

- The node can continue in service by updating its own copy of the database and by queuing its changes to the replication engine. Likewise, other database nodes in the network will be queuing their changes to the isolated node.

292

When the replication channel is restored, changes will flow in each direction to synchronize the databases. However, there are bound to be data collisions. The longer the node was isolated, the more collisions there will be. These must be detected and resolved, potentially a very large task.

- The node could be declared out of service and all users switched to their backup nodes as previously described.

Tug of War

Not to be forgotten is the tug of war problem that we discussed earlier. This occurs if a heartbeat network is lost, thereby leading two nodes to decide that the other is down. Both will attempt to seize the users serviced by the other node.

We have described several ways to prevent tug of war:

- Use an independent arbitration system.

- Have the heartbeat network and the user-switching network share the same channel.

- Switch users over manually.

Synchronous Replication

If synchronous replication or network transactions are being used, the presumption is that all nodes are operable. If a failed node does not respond to a commit request, the transaction is aborted. In effect, the entire system is down since no transactions can commit.

Should a node fail in this environment, it must be removed from the scope of the transaction. Its database must then be recovered following its return to service.

If queue draining is to be used, recovery in this case is a little different from that of a failed node under asynchronous replication. The failed node is first brought back into the network under asynchronous replication. Its database is then updated from the replication queue that had built during its outage (or from the audit trail or redo log if network transactions were being used). Once the replication queue has shrunk to a reasonable value, synchronous replication can be resumed.

Capacity Expansion

A new node can be integrated into the application network just as would a failed node following its recovery. Basically, it will first receive a copy of the application database. Its users will then be activated. Adding new nodes is discussed in Chapter 8, <u>Eliminating Planned Outages with Zero Downtime Migrations (ZDM)</u>.

What's Next

We have talked in the last three chapters about how to keep an active/active system operational even in the presence of failures. However, what about downtime that is imposed out of necessity, such as upgrades and capacity expansion? In the next chapter, we will talk about how the characteristics of an active/active system can be exploited to eliminate *planned* downtime with a technique known as Zero Downtime Migration.

Chapter 8 – Eliminating Planned Outages with Zero Downtime Migrations

"You just never know when some quirk of science or history is going to prove useful."

- Lynn Abbey

Introduction

Up to now in our discussions of active/active systems, we have concerned ourselves with surviving failures. The multinode architecture of active/active systems provides the opportunity to switch users over to a surviving node in the presence of a node or network failure.

However, if century uptimes are our goal, failure recovery is only part of the problem. In addition to these *unplanned* outages, today's systems and users are also burdened by *planned* outages. That is, systems must be brought down occasionally for all kinds of system maintenance. These planned downtimes are also included in the calculation of availability and must be eliminated if we are to achieve our availability goals.

There are many requirements that lead to planned outages:

- *Upgrades:* Vendor support for hardware, databases, and operating systems now has a definitive lifetime; and these components must be *upgraded* periodically in order to obtain continued maintenance and support.

- *Enhancements:* Competitive pressures require agility in our processing systems and lead to continual application *enhancement.*

- *Rearchitecting:* Technological pressures such as Web-enabling can lead to significant application *rearchitecting*.

- *Integration:* Mergers and acquisitions often require that diverse databases and applications be *integrated*.

- *Adoption:* Cost containment and the ever-changing availability of skills in the marketplace may mandate the *adoption* of emerging technologies.

- *Capacity Expansion:* The need for *capacity expansion* may require a migration to a larger system or to the addition of nodes in a distributed system.

- *Increased Availability: Increasing the availability* of a system in response to market expectations or customer demands may require migration to a fault-tolerant architecture or even to a distributed system that can tolerate node outages.

As a result, systems are necessarily undergoing frequent and significant changes. Hardware, operating systems, and databases must be upgraded to remain current and maintainable. Applications are continually being enhanced and ultimately rearchitected. Disparate systems must be integrated and new technology adopted. Migrations must be made to new systems to meet market demands, cost containment, or new availability goals.

In many existing systems, changes means denying service to users of critical applications, often for extended periods of time. Thus, the availability of mission-critical applications is seriously compromised.

Yes! Application Availability Counts

Availability is an ever-increasing requirement of today's systems. As companies go global with 24x7 customer service, and as costs of downtime escalate, increasing the availability of enterprise applications becomes a paramount requirement.

Upgrading a system for any reason often means that it will be out of service for a time that may be measured in hours, if not days. Prior to the Internet era, this may have been acceptable. Today it often is not. Migrations such as these must be done with zero application downtime. The applications must be continuously available to the system users with no "scheduled downtime."

Why are a few hours a year of downtime such a problem? Studies have shown that downtime of critical applications can cost anywhere from a few thousand dollars an hour to hundreds of thousands of dollars per hour to even the unthinkable - the loss of life or the business. A survey of enterprise users[91] showed that the average cost of downtime is between USD $10,000 and $100,000 per hour for over 50% of the respondents.

Even worse, downtime may have incalculable costs. A few minutes outage of a stock market's trading system will probably make the headlines. Such an outage in an emergency call system such as 911 service in the U.S. could result in a building burning down or a death due to cardiac arrest.

Of course, many critical applications such as an equities trading system are not 24x7. They have long periods of available downtime during which migrations or batch processing can be performed.[92] However, what happens if despite exhaustive testing during the available downtime, a serious bug is encountered during the first period of real operation? It may be imperative that users be returned

---

[91]Instapoll Archive, HP Online Advocacy Web Site (www.hpuseradvocacy.org); December 4, 2003.

[92] This may soon no longer be true, as many exchanges are going to electronic trading on a 24x7 basis.

to the original system immediately with no work being lost. If users cannot be returned immediately, they face *unscheduled* downtime.

The active/active system architecture can solve such challenges. It provides the facilities for bringing up a new system or a new environment with a database synchronized with the operational system, for testing that system with real users operating on live data and doing real work, for switching to the new system with no downtime, and for reversion to the original system with no work lost if there are problems. Scheduled downtime is eliminated, as is unscheduled downtime due to an upgrade or migration problem.

Thus, active/active systems can achieve *zero downtime migrations (ZDM)* from old applications and systems to new ones while solving the problem of scheduled downtime. We explore this very important capability further in this chapter.

Additionally, we will explore how the ZDM approach can be used to increase system capacity, load balance users across the system, and deploy an active/active system in the first place.

The ZDM Solution

An Overview

The ZDM solution is based on

- bringing up the new application as a separate instance from the operational application,

- synchronizing the new application's database with the operational application,

- testing the new application against a live database,

- optionally *live* testing or verifying the new application with an increasing number of real users for as long as needed (especially if external interfaces with no provision for testing are involved),

- reverting to the original application if necessary due to faults in the new application without loss of any transactions processed by the new application, and

- subsequent cutting over to the new application once it has been fully verified.

Online copy and bidirectional database synchronization facilities are central to this strategy. We have talked extensively about database synchronization using data replication engines. We will explore online copy facilities later in this chapter.

The ZDM Procedure

Let us look at the ZDM strategy, as illustrated in Figures 8-1a through 8-1c. Figure 8-1a begins with a two node active/active system running a current application with a current database. The system is symmetric – both systems are configured identically and are running the same application.

A need has arisen to modify the system. Perhaps the operating system has to be upgraded, or the database, or the application. Perhaps new hardware is going to replace the current hardware. Any combination of these upgrades may be required. The ZDM procedure is applicable to all of these needs.

The steps to be taken in the ZDM process are as follows:

Step 1: Isolate the Node to be Upgraded

One of the nodes is chosen to be the first one to be upgraded. The first task is to remove that node from the application network.

It is assumed that the remaining node is capable of carrying the entire application load. (If there are more than two nodes, the remaining set of nodes must be capable of carrying the entire application load.) Perhaps this has been accomplished by over-configuring the nodes. Alternatively, perhaps the upgrade is to be accomplished during slack times, during which the total load is significantly less than the configured peak load.

To remove a node from the application network, the first action is to switch all of the users that are connected to that node to the other node (or distribute them among multiple remaining nodes). Data replication to and from the node being isolated is then stopped.

Step 2: Upgrade the Isolated Node

The isolated node can now undergo whatever upgrades are required. It might be replaced with an entirely new system. The operating system, database, and applications are upgraded as necessary.

Step 3: Create a Test Database

It is now time to prepare the node for testing. This requires providing the node under test with a test database. This database is typically a copy (or a subset) of the actual current database. Alternatively, it may be a canned database used for regression testing.

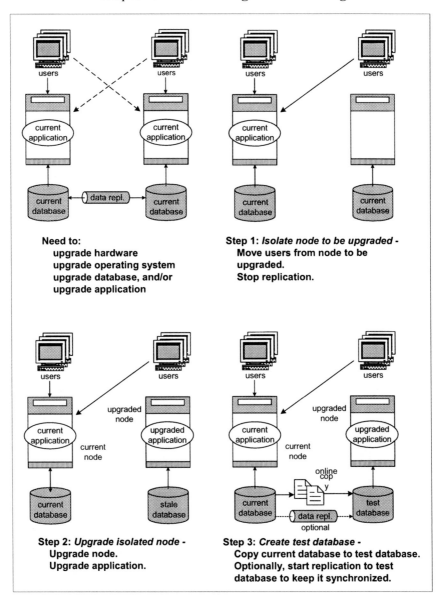

Zero Downtime Migration
Figure 8-1a

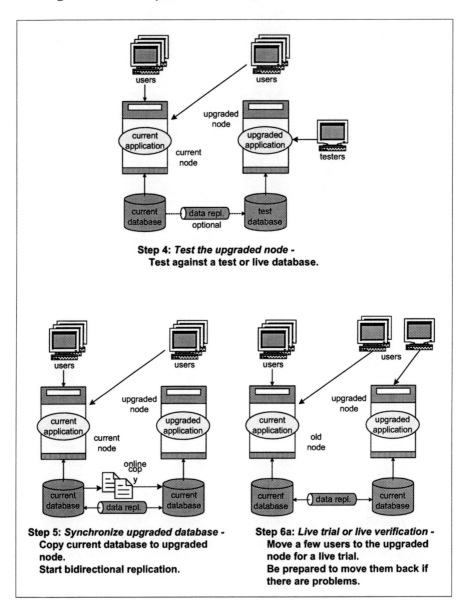

Step 4: *Test the upgraded node -*
Test against a test or live database.

Step 5: *Synchronize upgraded database -*
Copy current database to upgraded
node.
Start bidirectional replication.

Step 6a: *Live trial or live verification -*
Move a few users to the upgraded
node for a live trial.
Be prepared to move them back if
there are problems.

Zero Downtime Migration
Figure 8-1b

302

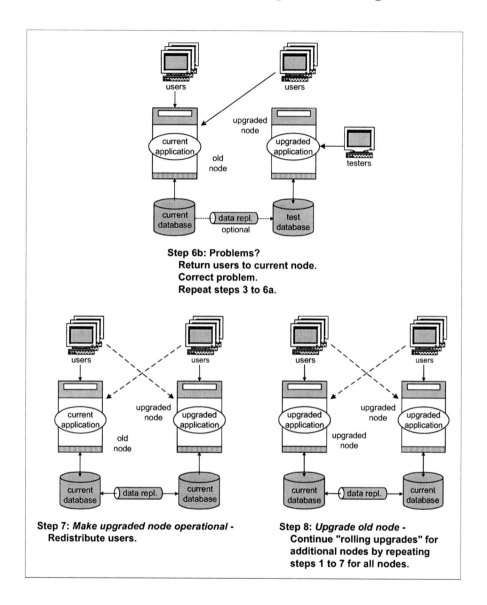

Zero Downtime Migration
Figure 8-1c

At this point, the node has been isolated from the network and can be taken down.

If the test database is to be a copy or a subset of the currently active database, that database or a subset thereof is copied to the database of the node to be tested. This must be done by a special copy procedure – an *online copy* – since the data that is being copied is also being actively updated by the online node or nodes.

The new database may have a totally different structure than the current database – for instance, it may be an SQL database, whereas the current database may be a nonrelational database. The database fields may be different. It is the job of the online copy utility to transform the data structures of the source database to the schema of the target database and to scrub and integrate the data as necessary.

All this must be done while the current application is actively processing transactions, with no downtime imposed on the applications or the users. The requirements for such an online copy facility are described later in this chapter.

If it is desirable to test or verify correct operations against a truly active current database, data replication may be started from the operational node(s) to the node under test. This is unidirectional replication since at this time, changes made to the test database should not be propagated to the rest of the application network. The replication engine is also responsible for data structure conversion, data scrubbing, and data integration.

Step 4: Test the Upgraded Node

Once the test database has been created, and once it is actively being synchronized with the current online database if desired, the testing of the upgraded node may begin. It can be put through normal testing for as long as is necessary because this testing can go on in parallel with the production application, all the while staying synchronized with the production data if necessary. At any point during the testing process, the new database can be purged and resynchronized with the production database in the event that it becomes corrupted as part of the testing process.

Up until now, the operational system and its users have been unaffected by the testing of the upgraded system.

Step 5: Synchronize the Upgraded Database

Once testing of the upgraded node has been satisfactorily completed, it is presumably ready to be put into service.

The first task to be accomplished is to ensure that the upgraded database is properly synchronized with the active database. This may require that the test database be purged and the online copy facility be once again invoked to copy the active database to the upgraded database.

At this point, bidirectional data replication can be started so that changes made in either database are reflected in the other. It is the responsibility of the data replicator to transform one data structure into the other as data is replicated in either direction.

The upgraded node is now ready for production use.

Step 6a: Live Trial or Verification

It is wise at this point to phase in the new node slowly to make sure that it is, in fact, working properly. Thus, this phase is really a live trial or verification.

A few users are first switched over to the upgraded node and are allowed to perform their normal duties. As this proves successful, more and more users can be switched over. This "ramp-up, ramp-down" strategy is especially important if the test facility is not fully featured or if external interfaces with rudimentary test capabilities are involved (for instance, if certain parts of the application's external interfaces can only by verified in a production environment).

In this mode, both databases are undergoing change with live transactions. Changes in either database must be replicated to the other using bidirectional replication to keep both databases synchronized. Assuming that asynchronous replication is being used, data collisions are now possible unless the database can be effectively partitioned between users for update purposes. Collisions can be detected by the replication engine and may possibly be resolved automatically, as described in Chapter 4, <u>Active/Active and Related Topologies</u>. Otherwise, they must be reported and resolved manually. Of course, if the system being upgraded is a node in an existing active/active network, the problem of data collisions is already being handled; and the collisions during this step have already been accounted for.

Step 6b: Problems?

Should problems be encountered during the ramp-up phase of the previous step, the users on the upgraded node can be switched back to the current node without losing any data; and the upgraded node can be corrected.

Steps 3 through 6a are then repeated to get the corrected upgraded node back on line.

Step 7: Make Upgraded Node Fully Operational

If the online trial or verification goes well, the upgraded node can be put fully into service by switching all of its users to it (or by phasing them over in groups).

Step 8: Upgrade Old Node

The application network is now running with one upgraded node and one old node. The bidirectional data replication engine is converting transactional data as necessary between the new and the old formats. The ZDM procedure, steps 1 through 7, is repeated to roll the upgrades to additional nodes so that the entire application network is operating with upgraded nodes.

Data Collisions in a Partitioned Environment

Partitioning a database is a common way to avoid collisions in an active/active environment. With a partitioned database, all updates to a particular data item are always made to the same database copy regardless of to which node the user is connected (although a user is often connected to the node that "owns" his data). These updates are routed over the network from the user's node and are applied to the appropriate database partition. They are then replicated to all other nodes.

However, during the ZDM process, there is a subtle race condition that could cause a data collision even if the database is partitioned. This occurs during the time that users are being switched from one system to another.

Consider a user who has just made a change to a partitioned data item, and is then immediately switched to another node. It is possible that he could make another change to that same data item in the new node, and that change beats the previous change to the database. The previous change will then arrive via replication and will overwrite the newer change. This is a data collision caused by the replication latency from the previous node to the new node.

Though unlikely (as is any data collision), it is more likely if the new node to which the user has been switched happens to be the node containing the database partition. The later update will be made directly to the database, while the earlier update must progress through the data replication channel. The faster the replication engine is, the less likely it is that this collision will occur.

One way to avoid this race condition is to make sure that the replication queue on the first node has drained prior to activating the user on the new node. This could be done by sending a special "marker" transaction to the new node through the data replication engine and by making sure that it has been posted prior to activating the users. This could be automated.

An easier way, perhaps, is to simply wait a short period of time to allow the replication queue to drain before activating the users on the new node. If the replication latency is subsecond or a very few seconds, this might be shorter than the time that it takes to switch the users anyway.

Of course, if the application expects collisions to occur, this is not a problem since they will be handled by the data replication engine's collision detection and resolution means.

Rolling Upgrades

As mentioned previously, the ZDM technique can be applied to more than two nodes that need upgrading, as shown in Figure 8-2. This is called *rolling upgrades*. First, one node is upgraded using the

ZDM methodology. This node is then used to upgrade the next node, which upgrades the next node, and so on until all nodes have been upgraded.

Thus, the fact that an application is distributed across multiple cooperating nodes in an active/active application network allows one node to be removed from the network for upgrading. Switching the users on that node to other nodes in the network results in the continuous provision of application services to all users while the required upgrade is being installed and tested. This upgrade then can be rolled to the other nodes in the system in a similar manner. The migration to a new revision of the application and/or its environment has been achieved with no application downtime required.

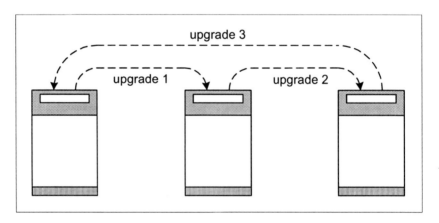

Rolling Upgrades
Figure 8-2

Single-System Upgrade

As mentioned earlier, it is important to emphasize that the ZDM technique is not dependent upon there being two separate physical nodes, as shown in Figure 8-3a. It is also applicable to upgrades within a single system, as shown in Figure 8-3b.

If, for instance, an application is to be upgraded on a monolithic system, and if this upgrade requires a change in the database structures, the new application can be brought up on the same system, as shown in Figure 8-3b. The above procedures then apply. The new database is created on the system's disk farm and is synchronized with the original database. The new application and database are then tested and subsequently put into service. Both databases can be kept synchronized via data replication so that reversion is possible if faults show up later in the upgraded system. This assumes, of course, that sufficient system resources such as disk are available to support the upgrade.

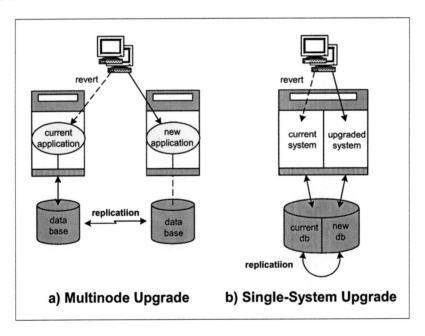

a) Multinode Upgrade   **b) Single-System Upgrade**

Single-System/Multinode Upgrades
Figure 8-3

Migrating to Active/Active

So far, we have talked about upgrading an existing active/active network. This, of course, assumes that we are currently using active/active technology. But perhaps a more complex problem is how we migrate an existing application into an active/active configuration in the first place.

There are actually two distinct steps in this process: making the application migration-ready and subsequently migrating it to an active/active configuration.

Making an Application Migration-Ready

Within this step, there are actually two substeps: (1) ensuring that the application can run in an active/active environment and (2) decomposing it so that it can be migrated.

Making an Application Active/Active-Ready

There are some applications that probably cannot be made to work in an active/active environment. For instance, if all incoming events must be processed in exact sequence (such as train movements along an instrumented track), these must all be processed by the same node; and true active/active cooperative processing is not possible.

Another example is a financial application that cannot tolerate data collisions. If synchronous replication is not available or is unsuitable due to application latency, and if the users cannot be partitioned to avoid collisions, then an active/active configuration may not be feasible.

However, active/active technology still can be used to create a "sizzling hot takeover" configuration, in which data replication to the backup node keeps it synchronized and ready to take over instantly in the event of the failure of the primary node (see Chapter 4, <u>Active/Active and Related Topologies</u>).

However, in the more common case of transaction processing systems, active/active is an achievable goal. Even here, there may be modifications that must be made to the application before it can be run in an active/active mode. Dr. Werner Alexi gives an excellent discourse on this topic in Appendix 4, <u>A Consultant's Critique</u>.

As he points out, these problems are generally related to shared global resources which must now be available to applications across the network. Examples include:

Shared Locks: To avoid deadlocks, a defined intelligent locking protocol (ILP) is often followed by each application. This protocol dictates in which order locks are to be acquired on the needed resources to avoid deadlocks. If the applications are distributed, they cannot simply acquire local locks on data since an application on another node may already have locked one or more of those data items on its copy of the data.

Instead, a global lock resource must be established that all applications must access in order to acquire locks. This might be implemented as a requirement to always lock a data item on a designated master node.

Unique Number Generators: Often, customer numbers or other identifiers are assigned by incrementing a unique number field. If this number is a data item in the database, it is globally accessible (though collisions could occur). If it is memory-resident, it must be partitioned somehow. This could be done by preceding the number with a node id, by assigning groups of numbers to each node, or by assigning specific sequences to each node (for instance, in a two-node system, node A assigns sequential even numbers and node B assigns sequential odd numbers).

Random Number Generators: A similar problem is that of a random number generator. If these numbers are calculated in different nodes, there is a distinct possibility that there will be duplicates. Again, this could be solved by incorporating a node number into each generated random number.

Transaction Distribution and Routing: In a multinode system, there must be some way to determine to which node a transaction should be routed. A simple solution is to assign each user to a specific node, with a backup node also assigned.

However, there are more sophisticated routing algorithms that might be advantageous, such as:

- round robin – send a transaction to the next node in sequence.

- load balancing – send a transaction to the least loaded node.

- content distribution – route the transaction according to its content, such as to the node containing the partition to be updated.

Transaction routing may be done either by intelligent network routers or by front-end processes in each node.

Care must be taken in partitioned databases since it is possible that a single transaction may update data items in several partitions. If the partitions cannot be assigned to avoid this,[93] the transaction must be broken into subtransactions directed to the appropriate partitions.

---

[93] Adding customer identifiers to keys and assigning customers to specific primary and backup partitions is a typical way to partition data to avoid collisions.

Local Context: There may be cases where context is contained not in the database but in memory. In this case, the context may not be globally accessible. A common example is connection context. If a transaction is sent to an external system which returns its response asynchronously on another connection, there may be no way to determine to which node the response should be returned.

Batch Processing: Batches are expected to be processed in only one node. However, there may be application decisions based on the current batch-processing status. Also, there may be mini-batches that are run under certain conditions that are global in nature, such as every 1,000 transactions.

Application Management: The application-management facility that monitors, configures, and controls processes will likely have to be upgraded to a network-aware facility.

All of these problems are solvable. It is simply important to be aware of these and other global problems and to modify the applications to behave properly in a distributed environment.

Decomposing an Application for Migration

Simply making the application active/active-ready may not be a sufficient condition for migrating to an active/active environment. The key is that the application and the database must be independent so that data replication can be used. Fortunately, most modern-day systems layer the application and the database so that independence is achieved.

However, in some older applications, especially those using a file-based database, the knowledge of the database schema may be buried in the application code. This makes the conversion and integration tasks of the data replication engine much more difficult.

Migrating such an outdated application to an active/active environment is no different than migrating it to a different platform. Such legacy migrations have been studied for years. An interesting discussion of this subject is found in Brodie and Stonebraker's book, *Migrating Legacy Systems: Gateways, Interfaces, and the Incremental Approach.*[94]

Brodie and Stonebraker argue that such migrations are so complex and risk-prone that they must be done incrementally (the Chicken Little approach) rather than as one big bang (the Cold Turkey approach). However, to migrate incrementally, the application must be *decomposable.* This means that the application must be independent of the database and the user and external system interfaces so that it can be broken up, or decomposed, into manageable modules. The modules can then be incrementally migrated.

Brodie and Stonebraker classify systems into three categories, as shown in Figure 8-4: decomposable, semidecomposable, and nondecomposable.

A decomposable system (Figure 8-4a) is one in which the user interfaces, applications, and database services are each independent of the others. A decomposable system is a good candidate for migrating to an active/active configuration.

A semidecomposable system (Figure 8-4b) is one in which the database services are partially or wholly embedded in the application. A nondecomposable system (Figure 8-4c) is one in which the user

---

[94] M. L. Brodie, M. Stonebraker, *Migrating Legacy Systems: Gateways, Interfaces, & the Incremental Approach,* Morgan Kaufmann Publishers; 1995.

interfaces, applications, and database services are all contained in one big monolithic structure.

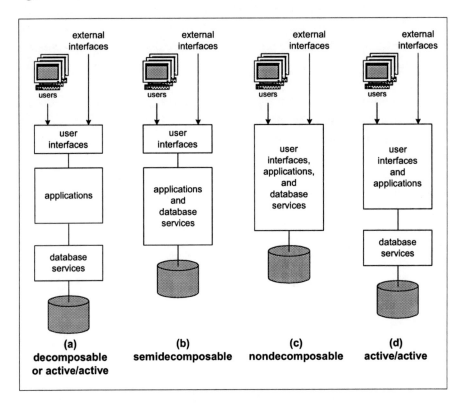

Decomposable Architectures
Figure 8-4

Actually, there is one additional configuration that is appropriate for active/active systems (Figure 8-4d). In these systems, the database is decoupled from the rest of the application, though the user interface is integrated into the application.

To be a candidate for an active/active system, the database services must be independent of the applications so that data replication can be used to synchronize the database with other database copies (Figures 8a and 8d). Thus, prior to migrating to an active/active configuration, database independence must be ensured. For newer systems, this is usually already the case. For older systems, this may entail some extensive application modification.

Putting the Active/Active-Ready System into Service

If major changes are required before the system can be used as the initial node in an active/active network, the modified system must be put into service as a standalone system and thoroughly tested in actual use.

If no changes were needed or the changes required to make the application active/active-ready are relatively minor (such as the changes required to correct the problems previously discussed in the section entitled Making an Application Active/Active-Ready), they can be incorporated over time via the normal system upgrade process. When these modifications are complete, the active/active-ready system is, in fact, in service and is ready to enter an active/active environment.

Alternatively, the current system could remain unmodified and the active/active-ready system could be brought up as a companion active/active node. A few users could be switched over as a trial, with reversion if necessary to correct problems. As confidence in the new system is gained, it could take on more and more users. Ultimately, all users would be switched to it and the old node upgraded via ZDM.

However, if the changes were of a fundamental nature, such as new hardware with a new operating system and database management system and/or a rewritten application, the modified system must first be put into service. This becomes a full migration effort, which is out of the scope of this book. Reference is made to Brodie and Stonebraker for an interesting discussion of this topic. Basically, they discuss the Chicken Little approach to legacy migration by using *gateways* to support incremental migration.

During the migration process, old and new interfaces must access old and new applications. Interface gateways provide translation from old and new interfaces to old and new applications.

Likewise, old and new applications must be able to access data in the old and new databases. Database gateways provide translation from old and new applications to old and new databases.

When all modules have been migrated to the new system, the active/active-ready system is ready to enter an active/active network as the initial node.

Migrating to an Active/Active Architecture

Once the active/active-ready system has been thoroughly tested, the stage is set for migration to an active/active architecture. This is done via the ZDM process, described in detail previously. A second system is brought up as a second node in the active/active network, and its database is synchronized with the current system. Bidirectional replication is turned on, and some users are switched from the original node to the second node.

It is important to treat this initial phase as a trial run or verification with only a few users, as it is likely that some problems that were not anticipated with concurrent operation may occur. If these are serious enough, the trial or verification users can be switched back to the original node while the problems are corrected.

As more and more users are switched to the new node, the system eventually becomes properly load balanced; and the new active/active network is up and running.

Uses for ZDM

ZDM has been described so far as a technique for doing system upgrades with no application outages. However, there are several uses for the ZDM process beyond this.

System Upgrades

The use of ZDM to upgrade the nodes in a system without denying service to any user has already been discussed in detail. See the section entitled The ZDM Solution to review this procedure. This section also applies to the other uses for ZDM, as discussed below.

Incremental Migration

Sometimes an upgrade is more complex than just a straight-forward upgrade to system hardware, applications, or the operating system or database management system. In these cases, incremental migration might be considered. Incremental migration could either be done at the application level or at the database level.

For example, a financial transaction processing company wanted to partition their database by customer. To do so, they added a new node to the application network and replicated a subset of customer data to the new node. They modified their applications that had to access this data to be server-aware. That is, each application was modified so that it could determine which node held the primary partition of the data for a particular transaction and could route the transaction to that node.

Bidirectional replication was then used to keep copies of the data partitions at each node for fast failback should it be required. Initially, only a few customers were moved over. If a problem occurred, the company would return to the unpartitioned system until the problem was corrected. As confidence was gained in the partitioning, more customers were moved over until the database was fully partitioned.

Note that if a partitioned application network has more than two nodes, significant database cost savings can be achieved by having each node contain only its partition and a copy of one other partition in the network. Each node is then backed up by one other node, with the consequence that only two database copies are provided for the entire system.

A similar need is solved with ZDM by a bank. In this case, the staff can not fully test their applications since these applications interface with many external systems. Although these external systems provide test facilities, these test facilities do not handle full load conditions or all of the anomalies that might occur in production usage.

To help alleviate this situation, the bank tests new applications as best it can. The staff then deploys them into production using a phased cutover. They partition the database by key range and copy that data to the new node to avoid collisions. Intelligent routers direct all traffic for a subset of data to the appropriate node for processing.

Ultimately, the bank switches all users to the new node, a process which can take weeks to months. When this is completed, the bank upgrades the old node and rebalances the users.

In addition to migration by incremental partitions, one could migrate by application. This can be used if the overall application is decomposable. Specific applications are migrated one or a few at a time in a controlled manner, as shown in Figure 8-5, so that the risk of migration failure is small.

As shown in Figure 8-5, not all of the applications need to be migrated as active/active sets. Only those that are critical need to run active/active. In Figure 8-5b, only Application 2 has been migrated in an active/active configuration. Other applications such as Application 3 in Figure 8-5b can be migrated with ZDM; and when confidence has been established in these, replication can be turned off and the application terminated on the source system.

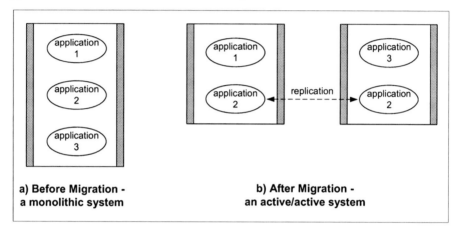

a) Before Migration -
a monolithic system

b) After Migration -
an active/active system

Incremental Migration
Figure 8-5

Capacity Expansion

Capacity expansion is conceptually easy with ZDM. If it is desired to add a new node to increase system capacity (or perhaps to take advantage of expanded locality), all that is required is to bring up that node with its applications. The node's database is then synchronized with one of the existing database copies, bidirectional replication is turned on to all other nodes, and users are switched over.

As mentioned earlier, if data partitioning is used to avoid collisions, one should ensure that all current transactions have drained out of the replication queue before activating the users on the new node.

Load Balancing

Active/active systems can easily be load balanced by simply redistributing users among the nodes. Load balancing is easiest if the node to which a user is connected is of no consequence. All that is required is that a subset of users be switched from a heavily used node to a lightly loaded node.

If data partitioning is used, and if users are associated with a partition (such as in the example given previously), it is necessary to move the data partition (or the routing table defining the new partitioning) along with the users. The technique for doing this was described above in the discussion concerning incremental migration.

The Online Copy Facility

Requirements

As we have seen, an important requirement for ZDM is the ability to copy part or all of the operational database to a test node that has been upgraded with new hardware, new applications, a new database, and/or a new operating system. This copy process should ideally be done while the database being created is also being actively updated so that stale data that has been previously copied is kept current. Otherwise, the subsequent replication of changes that have accumulated during the copy process could take a very long time.

The ideal online copy facility provides the following features:

- *Heterogeneous*: The source and target hardware, operating systems, and/or databases may be different.

- *High performance*: The time to copy very large databases should be as short as possible.

- *Noninvasive*: The application does not have to be modified in order to use it.

- *Low overhead*: The copy facility imposes a minimal footprint on the source system, and this overhead can be managed (throttled) by controlling the copy rate.

- *Target data transformations*: Source data can be reformatted or modified prior to being applied to the target database. The copy utility should use the same routines as the online data replication engine to ensure that copy and replication transformations are identical.

- *Data synchronization*: The copy facility cooperates with the data replication engine so that data synchronization occurs concurrently with the data copy. There will be no long queue of changes to be stored and replayed after the copy completes.

- *Consistency*: That part of the target database that has been copied is immediately in a consistent state, and its consistency is maintained. It may be used for application processing if appropriate.

- *Data compression, encryption*: Data compression and/or encryption is supported.

- *No downtime*: The application does not have to be paused during copying.

- *No disk utilization*: The copy facility does not require that a snapshot be taken and written to disk. It does not use any intermediate disk storage of the data to be loaded.

323

An example of a commercially available online copy facility that meets these requirements is described in Chapter 12 of Volume 3, SOLV.

Fuzzy Copying

An alternative approach (though problematic for a variety of reasons described below) to an online copy is what is known as a fuzzy copy. With this technique, data replication is paused while a shared extract (such as an SQL copy or load) is taken of the source database. This means that the source database is copied to the target database while the source database is being actively used and updated by applications. Therefore, the target database will not be in a consistent state during the copy process as rows copied later will be in a more recent state of update than rows copied earlier.

Once the shared extract has completed, this inconsistency is corrected by turning on replication and then draining the replication queue. When the replication queue reaches the same point as where the shared copy ended, the target database can be considered to be consistent. Thereafter, when the replication queue size reaches an acceptable level, the copy can be considered complete The target database will be synchronized but will be behind the source database by the replication latency.

Fuzzy copying has the following issues that must be recognized:

- It may take much longer than an online copy since the copy must be followed by applying the replication updates that have been queued. Draining the replication queue can take a considerable amount of time.

- It may use considerable disk space to store the replication queue that accumulates while the copy takes place.

- Many of the replication updates will be stale compared to the current state of the row or record which they are updating. Therefore, "fuzzy replication" must be used to update the target copy. If certain replication operations fail, they are mapped to other operations and retried. For example:

  - if an insert fails because the record already exists, the insert is converted to an update and is applied to the row.

  - if an update fails because the record does not exist, the update is converted to an insert and is added to the database.

  - if a delete fails because the record does not exist, the operation is ignored; or perhaps a row or record is reinserted with the relative difference.

Zero Downtime Migrations with Low Risk

There are two big problems with standard migration techniques:

- The current application must be taken down, perhaps for hours at a time, while an up-to-date database is created on the new system, while the new applications are brought up, and while users are switched over.

- Once the above is done, there is a leap of faith that the new application will work properly. This is the "big bang" syndrome – either it works, or the application services are down and are unavailable to the users. In this case, failback to the original system is required and may involve data loss.

ZDM solves these problems:

325

- There is no "big bang" cutover. The new environment is built, synchronized, and tested or verified without affecting the current users or production applications. Users will be phased over to a system that has been thoroughly tested in live operation.

- If desired, bidirectional replication allows the current system to be in service while the new system is integrated into the live environment. Both databases are kept in synchronism with each other via bidirectional replication.

- If there is a problem following cutover, users can immediately revert with no data loss to the original, known-to-be correct system, which has a completely up-to-date copy of the database.

- Users are never denied application services (except, perhaps, for a few seconds while they are being switched between applications).

- The original and upgraded system may use different hardware, different operating systems, different databases, different database structures, and/or different versions of the application.

- During online copying and resynchronization, full consistency is maintained at the target database for that portion of the database that has already been copied.

- Full support is provided via transformation facilities for the transformation of data applied to the target database.

Planned Outages Eliminated

A large amount of system downtime is due to planned outages required for upgrades, conversions, or migrations. The requirement might be for a new operating system version, an improved application, an enhanced database, or new hardware, among others. With ZDM, these can all be done without denial of application services to users. Scheduled downtime is eliminated.

> **Rule 53:** *ZDM (zero downtime migrations) eliminates planned application downtime, therefore improving application availability.*

What's Next?

By now, it is clear that active/active systems can bring extreme availability to applications. However, what do they cost; and is their cost penalty worth it?

In the next chapter, we explore the various cost issues associated with active/active systems so that a cost/benefit analysis can be accurately made.

Chapter 9 – Total Cost of Ownership (TCO)

"An investment in knowledge pays the best interest."
- Benjamin Franklin

As we've said in earlier chapters, you can optimize availability, cost, and performance. Pick any two. In Chapter 1, <u>Achieving Century Uptimes</u>, we reviewed the redundancy requirements necessary to achieve extreme a*vailability*. In Chapter 10 of Volume 3, <u>Data Replication Performance</u>, we explore *performance* issues. In this chapter, we discuss the *cost* of active/active systems.

Choosing the Solution

Before we can talk about the cost of a system, we must have determined the system that we are going to use. While this seems like an obvious statement, more often than not there are decisions to be made relative both to the details of the system configuration and how to pay for it.

In choosing a system and its specific configuration, Jim Johnson of The Standish Group makes an obvious but often ignored statement, which we paraphrase as follows:

Rule 54: (Standish Law) - *In order to calculate a meaningful predicted Total Cost of Ownership, one must first properly size the system.*

Put another way, you must properly size the system needed to do the job and not use some salesman's idea of a smaller system that will be unable to carry the load. Salesmen often have an incentive to undersize their systems (or to inflate their capabilities) and to oversize those of their competition (or to deflate their capabilities) in order to make the sale.

Different systems and system configurations will have different initial costs and different ongoing costs. Even given a specific system, there are usually many ways for which the system can be paid – high up-front costs with low monthly costs, low up-front costs with higher monthly costs, and so on. There may be additional options to consider, such as a distributed staff or a centralized staff with remote system management. There may be a variety of communication options to provide the required redundancy. There may even be a question of how to balance the cost of redundancy against the cost of downtime.

Each of these presents a different cost scenario. How is one to make a decision as to which is truly the least expensive approach, especially if the system is to be in operation for years? Up-front investment precludes income on that investment; this must be considered. Longer term payouts might cost more in total but may or may not cost more in reality when interest and investment loss is concerned.

The tools that can be used to make this decision are the net present value (NPV) and the internal rate of return (IRR).

Net Present Value (NPV)

The present value of an investment is that which would have to be invested (or spent) today to gain the same amount of income (or incur the same amount of cost) as a period of investments and revenues over a period in the future. It can be determined as follows.

Assume that an investment of PV dollars is made today and held for N periods[95] (a period is typically a month or a year) at a compounded interest rate of *i* percent per period.[96] PV is defined as the *present value* of the investment at the time that it was made. After the first period, the investment will be worth PV(1+*i*) dollars. After the second period, it will be worth PV(1+*i*)^2 dollars, and so on. After N periods, its value is PV(1+*i*)^N dollars. This is called the *future value* FV of the investment:

$$FV = PV(1+i)^N$$
(9-1)

What we are interested in is the present value of an investment or cost made in the future. This is, from Equation (9-1),

$$PV = \frac{FV}{(1+i)^N}$$
(9-2)

That is to say, spending FV dollars at the end of the Nth period in the future is equivalent to spending PV dollars today at the beginning of the period. PV is always less than FV because we can invest it and make money on it before we have to spend it in the future.[97] The same holds true for money received in the future. If we received it today, we could invest it and make money on it before the future date.

---

[95] An investment is made at the beginning of a period and returned at the end of a period.

[96] The interest rate *i* is actually more complicated. For an investment, it is the expected return. For an expenditure, it is the cost of money (what a bank will charge). Furthermore, *i* may vary over time; and inflation must be considered.

[97] This, of course, applies only to positive interest rates. There have been times, during war for example, when banks actually charged a fee to hold assets – in effect, a negative interest rate.

Therefore, when comparing the cost of different approaches, it is the total present value of the cost that should be considered and not simply the outlay of cash. The approach with the lowest present value cost is the more economical alternative.

Generally, not all system costs are made at the same time. We will make some investments now and some in the future. In this case, the present value of the system's total cost is the sum of the component present values of each cost, regardless of when that investment is made. This sum is called the *Net Present Value.*

Consider the following example. For a certain system, there are two system licensing options available for a ten-year period:

- <u>Option A</u>:   A one-time license charge (OLC) of $100,000 is due at the beginning of the first year. This covers the first eight years of licensing and maintenance. In the ninth and tenth years, an annual service charge (ASC) of $28,000, due at the beginning of the year, is charged.

- <u>Option B</u>:   An OLC of $40,000 is due at the beginning of the first year. Thereafter, for four years, an annual licensing and service charge (ALSC) of $28,000 per year is due at the beginning of each year. Licensing and service are then paid up for the remaining five years.

Option A's total cash outlay (or fees) is $156,000 over the life of the contract, and Option B's total fees are $152,000. However, which is really the most attractive? One way to make this determination is to calculate the net present value of each approach. This is done in the following Table 9-1 and presumes an 8% interest rate.

Option A		Year	Option B	
Cost	PV		Cost	PV
100,000	100,000	1	40,000	40,000
0	0	2	28,000	25,927
0	0	3	28,000	24,005
0	0	4	28,000	22,227
0	0	5	28,000	20,581
0	0	6	0	0
0	0	7	0	0
0	0	8	0	0
28,000	15,128	9	0	0
28,000	14,006	10	0	0
156,000	129,134	Total	152,000	132,740

**Net Present Value of Options A and B at 8% Interest
Table 9-1**

It is interesting to note in this example that Option A has the lower net present value cost ($129,134) and therefore is less expensive than Option B ($132,740) even though Option A's total fees ($156,000) are more than those of Option B ($152,000). The least cost alternative is not always obvious.

Internal Rate of Return (IRR)

In the above example, it was shown that by considering the net present value, Option A was the less expensive of the two options for an interest rate of 8%. However, if the interest rate is 2% rather than 8%, then the above analysis would show that Option B is the least costly option.

This raises the following question: At what interest rate do the two options have the same net present value? For interest rates less than the breakeven interest rate, Option B is preferred. For interest rates greater than the breakeven interest rate, then Option A is less costly.

This is equivalent to asking what interest rate makes the net present value of the difference between the two approaches zero. The effective interest rate of an investment is called the internal rate of return (IRR). The IRR is the interest rate received for an investment consisting of payments (negative values) and income (positive values) that occur over various periods. If we can determine the interest rate that gives a zero net present value to the difference between the two approaches (i.e., a zero return), this "breakeven" interest rate can be used to choose the best option. We will call this IRR_0.

Let us illustrate the above by calculating the IRR_0 breakeven interest rate for Options A and B. We first determine the differences over the periods for the two approaches:

- The initial cost for Option A is $60,000 more than Option B in year 1.
- Option B is $28,000 more in years 2 through 5.
- Option A is $28,000 more in years 9 and 10.

As reflected in Table 9-2, the difference is taken as negative if it represents a reduction in costs (an income of sorts) for Option A and positive if it represents an increased cost for Option A (an investment of sorts):[98]

---

[98] It doesn't make any difference which sign convention is used as the breakeven interest is the same in either case.

Year	Option A	Option B	Difference
1	100,000	40,000	60,000
2	0	28,000	-28,000
3	0	28,000	-28,000
4	0	28,000	-28,000
5	0	28,000	-28,000
6	0	0	0
7	0	0	0
8	0	0	0
9	28,000	0	28,000
10	28,000	0	28,000
Total	156,000	152,000	4,000

Differences between Options A and B for IRR
Table 9-2

For this example, the net present value of the difference is found by using Equation (9-2) to add the present values of the components of the difference The breakeven interest rate IRR_0 is that which will make this expression zero:

$$60,000 - \frac{28,000}{(1+i)} - \frac{28,000}{(1+i)^2} - \frac{28,000}{(1+i)^3} - \frac{28,000}{(1+i)^4} + \frac{28,000}{(1+i)^9} + \frac{28,000}{(1+i)^{10}} = 0$$

Unfortunately, the breakeven interest rate, i, that satisfies this relationship must be determined iteratively. There are calculators which will do this, including Microsoft's Excel function IRR.[99] Using that function, the breakeven value for the interest rate, in this case, is

---

[99] The IRR function in Excel handles only a fixed interest rate for fixed periods. Excel itself can be used to calculate more complex situations.

2.55%. That means that if the cost of money (or return on investment) were 2.55%, the net present value cost of both approaches would be the same ($145,198 in each case, to be exact).

Now that we know the breakeven interest rate, which option, A or B, is best for some real interest rate? Let us begin by looking at the net present value for each option at an interest rate $i = 0$. From Equation (9-2), we can see that if $i = 0$, the net present value is the same as the future value (the actual cost of the licenses). Let us represent this as NPV(0). As we can see from the above Table (9-2), the zero-interest NPV for Option A is $NPV_A(0) = \$156,000$, and for Option B is $NPV_B(0) = \$152,000$. Therefore, if the real interest rate to be paid is zero, Option B is the better choice.

However, what about other interest rates? Since net present value gets smaller as i increases, as shown in Figure 9-1 (see also Equation (9-2)), it is clear that for any value of the breakeven interest rate, IRR_0, if the real interest rate to be paid is greater than IRR_0, the option with the larger NPV(0) will be the best choice (Option A in this case). If the real interest rate is less than IRR_0, the option with the smaller NPV(0) will be better (Option B in this example).

This rule is summarized below:

- If the real interest rate to be paid is less than IRR_0, choose the option with the smaller NPV(0).
- If the real interest rate to be paid is greater than IRR_0, choose the option with the larger NPV(0).
- If the real interest rate is equal to IRR_0, both approaches are equivalent.

Of course, life is never so simple. In actual practice, the cost of money (what it costs to borrow) may well be different than the time value of money (what the funds can earn). The periodic costs of each approach may vary from period to period, and further investments may have to be made from time to time. Inflation may play a factor. Microsoft's Excel provides a function to calculate IRR with varying investments and costs. However, it does not handle additional complexities such as inflation and varying costs of money. In such cases, a more sophisticated calculator must be used; or a spreadsheet like Excel can be used to calculate the actual net present values and to compare them.

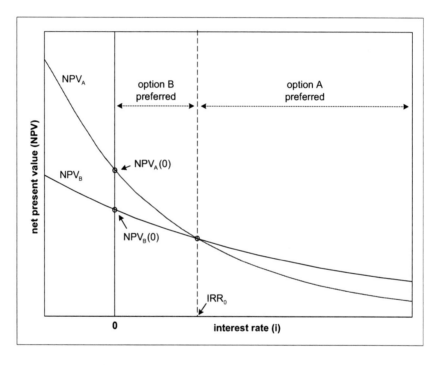

Choosing An Approach Via IRR
Figure 9-1

To formalize the general case in which investments are different for each period and in which interest rates are different for each investment for each period, consider the investments in a computer system as shown in Table 9-3.

Notice that there are several costs – hardware, software, facilities, personnel, and so on (there also could be some income, which we have called a return). There are a total of K cost categories. The kth cost is spread over N periods ranging from period 0 (the up-front costs) to period N. Each payment, except for period 0, is made at the end of the period. The future value of the kth cost for the pth period (that is, the investment during that period) is FV_{pk}.

		period (p)				
		0	1	2	3	. . . .	N
cost (k)	1. Hardware Lease	50,000	10,000	10,000	10,000		10,000
	2. Software Licenses		5,000	10,000	10,000		20,000
	3. Facilities		3,000	4,000	4,000		5,000
	. . . .						
	K. Personnel		18,000	22,000	25,000		30,000

Generalized NPV Example
Table 9-3

For each cost k and period p, there is an interest rate i_{pk}. (The cost for $p = 0$, the up-front cost, might be zero; or it might be points.) The present value PV_{pk} for the pth period of the kth cost is, from Equation (9-2),

$$PV_{pk} = FV_{pk} \prod_{j=0}^{p} \frac{1}{1+i_{jk}}$$

where j is a period index, and i_{0k} is typically zero (i.e., there is no interest applied to up-front costs).

The net present value for the kth cost, NPV_k, is the sum of the present values for this cost item over the N periods:

$$NPV_k = \sum_{p=0}^{N} PV_{pk} = \sum_{p=0}^{N} FV_{pk} \prod_{j=0}^{p} \frac{1}{1+i_{jk}}$$

The net present value for all costs, NPV, is the sum of the net present values of the K individual cost items:

$$NPV = \sum_{k=1}^{K} NPV_k = \sum_{k=1}^{K} \sum_{p=0}^{N} FV_{pk} \prod_{j=0}^{p} \frac{1}{1+i_{jk}} \quad (9\text{-}3)$$

where

FV_{pk} is the kth investment or return made or received at the end of the pth period ($p \geq 0$). Period $p = 0$ is an up-front investment or return.

i_{pk} is the interest rate for the kth investment or return during its pth period.

N is the total number of periods.

K is the number of individual cost items.

Return on Investment (ROI)

The driving force behind most enterprise investment decisions is Return on Investment (ROI). It is typically stated either as a rate of return or as the time in which the investment will pay for itself. For example, if a new system is to cost $1,000,000, will be paid up front, and will save the company $100,000 per year, then its ROI is 10%. It will pay for itself in ten years, assuming a 0% interest rate for simplicity. A more accurate payback calculation is obtained by using the net present values of the investments and returns rather than the future values as we did here.

There are two parts to the ROI equation – what will the investment return, and what will it cost. The first factor – what will the investment return – is very much a business process issue not explored here except for the cost of downtime. The second issue – what will the investment cost – is a subject explored further in this chapter.

Often, what a system will cost is in the eye of the beholder. If a manager is required to meet quarterly numbers, only the initial system costs may be of interest. If recurring costs are considered, some that are incurred anyway and that are seen as not changing may be excluded, such as facility costs.

In the rest of this chapter, we consider all potential costs, both initial and recurring. Over the life of the system, the sum of all costs adjusted for inflation and for the time value of money is called the Total Cost of Ownership (TCO). The TCO is, in fact, the net present value of the investment, as discussed previously. It is not the purpose of this chapter to present actual costs. Rather, we now will discuss those major cost items which should be considered in the TCO calculation.[100]

Total Cost of Ownership (TCO)

The focus of this book, and therefore the focus of this discussion on TCO, is that of extremely reliable systems – perhaps six 9s or better. Therefore, it is assumed that all components within the system are redundant, that the system is rapidly recoverable following a failure, and that redundant components are geographically distributed to protect the system from local disasters.

The TCO components which we will consider include:

- initial system cost
- maintenance
- downtime
- software

---

[100] For further information on TCO, see www.standishgroup.com. Their useful service, VirtualADVISOR Solution, is based on thousands of business case studies. VirtualAdvisor is described later in this chapter.

- communications
- network management
- personnel
- facilities
- business risk insurance

There are probably other costs that may be incurred by an organization in certain cases, but our list contains the most common significant costs.

Initial System Cost

The initial system cost is usually self-evident – just look at the price quotes or purchase orders. This cost includes the monies spent for the processors, disks, communication equipment, terminals, printers, and any other hardware components that must be purchased to implement the system as well as for up-front system and application software licenses or software development. It may also include the up-front cost of financing or the lost opportunity of cash that might otherwise be invested. Also,

Rule 55: *Remember the power of negotiation. For instance, at the end of a vendor's fiscal quarter, sales goals may drive significant discounts.*

For active/active systems, there are a variety of configurations which might be used, each with its own cost/availability tradeoff. We discuss these options next.

Processing Redundancy

Since we are concerned with extremely available systems, we are therefore dealing with redundant systems. Redundancy can range anywhere from an idle backup system to multiple nodes in an active/active system. Each of these represents a different

341

cost/availability tradeoff. We have also seen that the amount of processing hardware needed can be a function of the reliability of the network nodes. For example, systems comprising nodes with three 9s availability generally require more nodes than those systems comprising nodes with four 9s availability in order to obtain the same availability profile.

We have analyzed this in some detail in Chapter 1, <u>Achieving Century Uptimes</u>, and have summarized the results in Table 2-1 of Chapter 2, <u>Reliability of Distributed Computing Systems.</u> The results summarized in Table 2-1 are the basis for our discussion here, which is a review of the impact of those findings on TCO.

One fundamental choice to be made when determining initial system cost is whether a highly-available system or a fault-tolerant system will be used. Both comprise multiple nodes and are suitable for active/active systems. High-availability systems, characterized, for instance, by Unix clusters,[101] provide a backup for every component. If a component should fail, the backup component replaces it. This often requires that applications be brought up, that databases be opened, and that in-doubt transactions be recovered. This process usually takes in the order of minutes, or even hours if management approval is required, during which time the system is unavailable to the users affected by the failure.

A fault-tolerant system, characterized, for instance, by HP's NonStop servers,[102] is similar in that there is a backup for every component in the system. The difference is that applications are backed up by active applications, which remain synchronized with the primary application. If an application is taken down by either a hardware or software failure, the synchronized backup application

---

[101] Such as Microsoft Cluster Servers, Sun Clusters, or HP's ServiceGuard. See Peter S. Weygant, *Clusters for High Availability,* Hewlett-Packard Professional Books; 2001.
[102] See www.hp.com/go/nonstop.

takes over in a matter of seconds. The difference between a five-minute recovery (300 seconds) for a high availability system and a three-second recovery for a fault-tolerant system is two 9s of availability (i.e., downtimes are different by a factor of 100).

Figures 9-2 and Figure 9-3 show several examples illustrating different configurations that one might use to achieve extreme availabilities. They are similar to those configurations shown in Figures 1-4 and 1-5 in Chapter 1 with the addition of relative TCOs which consider, among others, initial cost, facilities, people, and downtime. Figure 9-2 represents sample configurations when high-availability nodes with three 9s availability (such as Unix or Windows systems) are used. Figure 9-3 shows similar systems if fault-tolerant systems with four 9s availability (such as NonStop servers) are used.

These diagrams show the approximate TCO relative to a single system. It is assumed that facilities costs, people costs, network costs, and maintenance costs are roughly proportional to initial system cost. Downtime cost is the cost of a failure multiplied by the probability of failure. It can be unlimited (e.g., loss of the company) for the more fault-prone systems (three to five nines – thus the notation "1.0 & ↑,") but is small for very reliable systems (six nines and above) due to the low probability of experiencing a fault.

Highly-Available Nodes

If highly-available nodes are used, such as Unix or Windows systems, we assume that they have an availability of three 9s. If the system is implemented as just a single node (Figure 9-2a), then it will have an availability of three 9s and can withstand no failures.

In a standard active/backup configuration (Figure 9-2b), the system can withstand a single node failure and will have an

343

availability of six 9s. However, a total of 200% of the required computing capacity must be purchased.

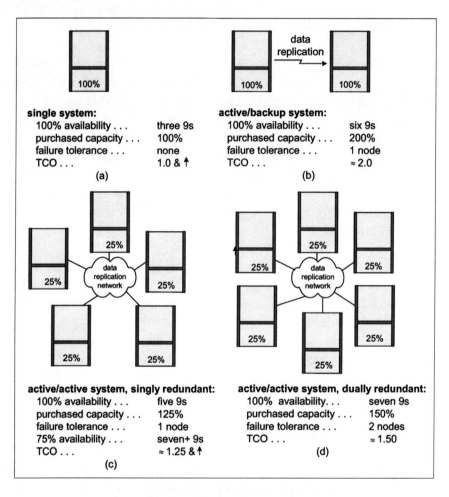

single system:
  100% availability . . .   three 9s
  purchased capacity . . .   100%
  failure tolerance . . .   none
  TCO . . .   1.0 & ↑
  (a)

active/backup system:
  100% availability . . .   six 9s
  purchased capacity . . .   200%
  failure tolerance . . .   1 node
  TCO . . .   ≈ 2.0
  (b)

active/active system, singly redundant:
  100% availability . . .   five 9s
  purchased capacity . . .   125%
  failure tolerance . . .   1 node
  75% availability . . .   seven+ 9s
  TCO . . .   ≈ 1.25 & ↑
  (c)

active/active system, dually redundant:
  100% availability. . .   seven 9s
  purchased capacity . . .   150%
  failure tolerance . . .   2 nodes
  TCO . . .   ≈ 1.50
  (d)

Relative TCO of Redundant Systems with Highly-Available Nodes
Figure 9-2

Figures 9-2c and 9-2d show active/active configurations comprising multiple smaller nodes, each of which can handle 25% of the capacity. Figure 9-2c shows a five-node system that can withstand a single node failure. It has an availability of five 9s. A total of 125% of processing capacity must be purchased.

Figure 9-2d shows that same configuration with six nodes. It can withstand a failure of two nodes and therefore has seven 9s of availability. 150% of processing capacity must be purchased.

Clearly, the only configurations that meet the extreme availability requirement of six 9s or better are the active/backup configuration (Figure 9-2b) and the dually redundant active/active system (Figure 9-2d). The active/active configuration requires 25% less purchased capacity (150% versus 200%) and shows the economics that active/active systems can provide.

All of these cases have assumed that in the event of a failure, 100% of capacity must still be provided. Significant economies can be achieved if this requirement can be relaxed. For instance, if failure is defined as providing less than 75% of capacity, then the five-node configuration of Figure 9-2c can withstand a two-node failure and will provide a little more than seven 9s of availability at 75% of capacity. Thus, we have achieved our goal of extreme availability with one less node – only 25% more capacity than normally required must be purchased.

Fault-Tolerant Nodes

Figure 9-3 shows the equivalent configurations using fault-tolerant nodes such as NonStop servers. A fault-tolerant node has the capability of surviving at least one hardware or software fault. Two

345

such faults may take down the node. It is this ability to survive a fault of any kind that gives it an order of magnitude better availability (typically, four 9s rather than three 9s).

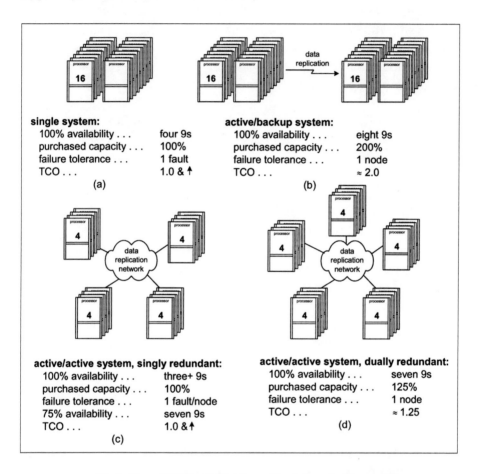

single system:
   100% availability . . .   four 9s
   purchased capacity . . .   100%
   failure tolerance . . .   1 fault
   TCO . . .   1.0 & ↑
            (a)

active/backup system:
   100% availability . . .   eight 9s
   purchased capacity . . .   200%
   failure tolerance . . .   1 node
   TCO . . .   ≈ 2.0
            (b)

active/active system, singly redundant:
   100% availability . . .   three+ 9s
   purchased capacity . . .   100%
   failure tolerance . . .   1 fault/node
   75% availability . . .   seven 9s
   TCO . . .   1.0 & ↑
            (c)

active/active system, dually redundant:
   100% availability . . .   seven 9s
   purchased capacity . . .   125%
   failure tolerance . . .   1 node
   TCO . . .   ≈ 1.25
            (d)

**Relative TCO of Redundant Systems with
Fault-Tolerant Nodes
Figure 9-3**

Figure 9-3 illustrates a 16-processor fault-tolerant system that is separated into four-processor nodes to provide an active/active architecture. We assume that a single system (Figure 9-3a) will provide 100% capacity with an availability of four 9s. In a standard active/backup configuration (Figure 9-3b), 100% capacity will be provided with eight 9s availability. However, 200% of the needed capacity must be purchased.

Figures 9-3c and 9-3d show this system configured in an active/active configuration with four-processor nodes that each provide 25% of full capacity. Four operational nodes are required in order to provide 100% capacity. We continue to assume that each node has an availability of four 9s.[103]

Figure 9-3c shows a four-node configuration. This configuration cannot withstand any node failures. Though it can survive one fault in each node, two faults in the same node could take down a node and thus the system. Interestingly enough, this configuration provides 100% capacity only with an availability of something more than three 9s (because of its four failure modes), not much better than a single high availability system. In this case, only 100% of the required system capacity has been purchased.

Figure 9-3d shows a five-node configuration of fault-tolerant systems. It can withstand one node failure and will provide seven 9s of availability. 125% of full capacity must be purchased.

For these examples, it is clear that only the active/backup configuration and the five-node active/active configuration will provide the desired availability. The active/active configuration requires the purchase of 125% of capacity compared to 200% for the active/backup configuration.

---

[103] Actually, there are reasons that smaller nodes have higher availabilities, as described in Chapter 2, Reliability of Distributed Computing Systems. However, we ignore this advantage for purposes of these examples.

Also, in a manner similar to that noted for high availability systems, if providing 75% of capacity following a failure is acceptable, then the four-node configuration of Figure 9-3c is suitable. It can withstand a one-node failure and will provide 75% of capacity with an availability of seven 9s. In this case, no additional capacity need be purchased. It is sufficient simply to reconfigure the monolithic system as an active/active system.

Comparison Summary

The comparisons of the above examples are summarized in Table 9-3, which gives the purchased capacity required to achieve the availability goals.

Configuration	Highly-Available Nodes	Fault-Tolerant Nodes
active/backup	200%	200%
active/active:		
100% availability	150%	125%
75% availability	125%	100%

Purchased Capacity for Six 9s
Table 9-3

From the table, one notable conclusion is apparent. In the above examples of active/active systems, a given level of availability is achieved with one less node if fault-tolerant systems are used. Achieving 100% capacity with an availability in excess of six 9s requires five fault-tolerant nodes versus six high availability nodes. Achieving 75% capacity with an availability exceeding six 9s requires four fault-tolerant nodes versus five high availability nodes. This observation was stated in our earlier Rule 39, formulated in Chapter 1:

Rule 39: (NonStop Maxim) - *Century uptimes in an active/active system will generally require fewer fault-tolerant nodes than required by an active/active system using high availability nodes.*

An additional observation is that any of the active/active configurations require significantly less purchased capacity than the classic active/backup configuration.

From Rule 39, it is clear that less capacity need be purchased if fault-tolerant nodes are used instead of highly-available nodes. But how much less? The examples shown in Figures (9-2) and (9-3) were based on splitting a system into four nodes. However, what if the system is split into two nodes? Into ten nodes? To determine the required purchased capacity, one must do an analysis similar to the above. Chapter 2 gives the mathematical tools to do this and results in the simple relationship (Equation (2-17)):

$$A \approx 1 - f(1-a)^{s+1}$$

(2-17)

where

A is the required availability of the system.
a is the availability of a node.
s is the number of spare nodes (i.e., the number of nodes that can fail without taking down the system).
f is the number of ways that a system failure can occur (the number of ways that $s+1$ nodes can fail – see Equation (2-15) in Chapter 2).[104]

---

[104] If all nodes are equivalent, then $f \leq \dfrac{n!}{(s+1)(n-s-1)!}$, where n is the total number of nodes in the system, including s spares.

We assume that the availability of a single node, *a*, is known. To size the system, the first step is to determine the number of nodes that are needed to support the application. Equation (2-17) is then solved for *s*, the number of additional spare nodes that must be purchased to achieve the desired availability *A*. (Excel is a handy tool to make this calculation.) The analysis will predict the required capacity that must be purchased either for highly-available nodes or for fault-tolerant nodes.

Just because less capacity must be purchased for fault-tolerant nodes does not mean that this approach will be less expensive. It depends upon system pricing, which is the final exercise in determining initial processing costs.

Storage Redundancy

Disk storage is often a major consideration in the initial cost of a system. It can represent more than half the cost if large disk farms are needed.

As discussed in Chapter 3, <u>Building Active/Active Systems</u>, there are a variety of options to configure disk subsystems in an active/active network. Each provides a different tradeoff between cost, availability, and performance. We review these configuration options here.

<u>Disk Configuration</u>

A fundamental option that cuts across all network configurations to be discussed is the disk configuration itself. One can use single disks, mirrored disks, or RAID (Redundant Arrays of Independent Disks[105]). These, in fact, can be intermixed and used in different areas

---

[105] Originally, Redundant Arrays of Inexpensive Disks.

in the active/active network. Each represents a different compromise between cost and availability (and in some cases performance).

Let us look at the availability characteristics of these three options.

Single Disk

Disk units today have become very reliable. A typical disk has an average time to failure of about 500,000 hours. Assuming a leisurely average repair time of 24 hours, such a disk has an availability of more than four 9s.

Mirrored Disks

Four+ 9s is good, but it is not adequate for the extreme availabilities of six 9s or better for which we are looking. A system typically has many disk drives. If it has ten disk drives, then the availability of the disk subsystem will only be three 9s. If it has 1,000 disks, the disk system availability is only one 9. Some very large systems exist with many thousands of disks.

Therefore, it is common practice today to mirror disks, providing a backup disk to take over in the event of the failure of the primary disk. Mirroring will yield a respectable availability of nine 9s for a single mirrored disk at about twice the cost. For large systems with thousands of disks, this still yields availabilities in the order of six 9s.

RAID

A powerful compromise between single and mirrored disks in terms of cost and availability is RAID.[106] In a RAID array, data is striped across multiple disks in such a way that any one disk can fail

---

[106] There are multiple levels of RAID arrays. Mirrored disks are technically RAID 1.

and the array will still have the data available. In effect, one extra disk is provided; and the data is arranged so that it can be reconstructed from any subset of disks should one fail. RAID arrays also can be structured to have more than one spare disk.

Let us take as an example an application which needs the storage capacity of five disks. A RAID array might be selected that stripes the data across six disks. Should any of these disks fail, the data is still available. It would take a dual disk failure to bring down the RAID array.

Assuming that our disk example described above is used, the availability of any two disks in the RAID array is nine 9s. However, in this case, there are fifteen ways in which two disks can fail in a six-disk RAID array (see Equation (2-17). Therefore, it will have an availability of more than seven 9s.

This availability lies between that of the single and mirrored disks, as does the cost of the RAID array.

Comparison

The comparison between disk configurations is shown in Figure 9-4 and is summarized in Table 9-4.

The storage network configurations in the next sections show the use of single and mirrored disks. All are configurations which meet the exacting requirements for extreme availability. However, in each case, a RAID array may be substituted for the disk subsystem. If the disk is a single disk, then RAID will make it more reliable at some additional expense. If the disk is mirrored, substituting RAID will reduce availability but will also reduce cost.

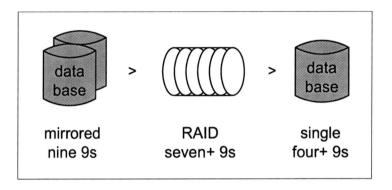

Disk Configuration Availabilities
Figure 9-4

	Availability	Purchased Capacity	Cost
Single disk	four+ 9s	1.0	least expensive
RAID array	seven+ 9s	1.2	↓
Mirrored disks	nine 9s	2.0	most expensive

Disk Subsystem Option Examples
Table 9-4

Configuration Options

Figure 9-5 shows a variety of storage network configurations suitable for active/active systems. They each are different compromises between cost, availability, and performance. As in Figures 9-3 and 9-4, illustrative TCOs are shown under the simplifying assumption that the cost of a simplex disk subsystem (i.e., the complement of disks to hold one copy of the application database) is comparable to that of a node. These options are explored next.

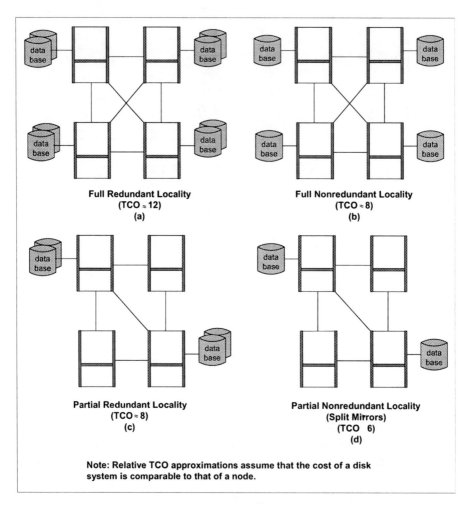

Active/Active Storage Configurations
Figure 9-5

Full Locality

Full Redundant Locality

Figure 9-5a illustrates the premier configuration. It has the highest availability, the highest performance, and, of course, the highest cost.

In this configuration, each node in the network has its own mirrored copy of the database and provides high availability of the database locally to each node. Because each node has local access to its data, performance is optimized.

Full Nonredundant Locality

A fully redundant system seems to have an overabundance of duplication. Why do we need multiple mirrored copies of data across the network? Is this not overkill?

In some cases, it may not be excessive, especially if locality of data is critical. Other times, multiple mirrored copies may be unnecessary. Certainly, in our example of a four-node active/active system, having a single copy of the database at each node as shown in Figure 9-5b still provides more database redundancy than a mirrored database at a single node.

If only a single copy of the database is to be provided at each node, we must decide what the procedure will be in the event of a single media failure which causes a database failure at one node. Since these are unmirrored disks, the availability of a single disk is similar to that of the node itself (three to four 9s). Since there are generally multiple disks per node, the failure of a single disk may be significantly more likely than the failure of the node.

355

In the event of a media failure, there are two options. The node can be considered to have failed and its users switched to surviving nodes. Alternatively, the node can be left up and a remote database used (see Chapter 5, <u>Distributed Databases</u>).

The latter choice could have serious performance consequences if the node is geographically distant from its surrogate database. This is due to the transit time over long-haul networks (about 20 milliseconds round trip per 1,000 miles). The transit time delay would affect each read and write operation from the remote node. However, if each node has another node with a surrogate database that is relatively close (say within 100 miles), performance degradation may not be apparent.

Partial Locality

Partial Redundant Locality

If the network topology is such that there are groups of geographically close collections of nodes, some nodes could share a database located at one of the other nodes in its group. Figure 9-5c shows such a configuration, in which any node has access to a mirrored database through one communication hop at most. The database nodes are themselves interconnected to support database synchronization.

Taking advantage of locality in this way can dramatically reduce disk costs.

Partial Nonredundant Locality

Again, one might question whether multiple mirrored disks are necessary in the network. Certainly, a single mirrored database provides acceptable availability in a single-node system.

Therefore, one might consider a partial locality system as shown in Figure 9-5d, in which each node that has a database has only a single copy – not mirrored. This configuration can be carried to the point that there are only two copies of the database in the entire network. This is called the split-mirror configuration and provides the same availability as a single mirrored-disk pair on a single system. In addition, it survives disasters if the nodes are geographically separated.

In this case, however, each node that does not have a local database must be given access to two database nodes since otherwise the failure of a single database at one node precludes the extreme availability which we are seeking.

Network Attached Storage (NAS)

Mirrored NAS

A radically different configuration from that of distributed databases is the use of network attached storage (NAS), as shown in Figure 9-6. With this technology, a self-standing storage system is available to any node in the active/active system via the network. Figure 9-6a shows the case of a mirrored NAS subsystem.

NAS is certainly an economical solution for distributed access to data in an active/active system. However, it may suffer from poor performance since all data accesses must be across the network in a geographically dispersed system.

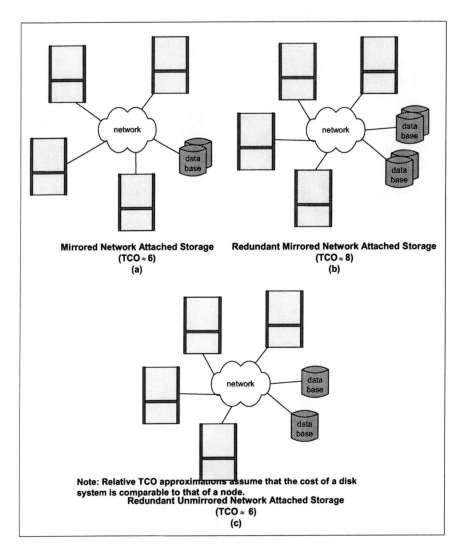

Network Attached Storage Configurations
Figure 9-6

Redundant Mirrored NAS

There is one problem with the configuration shown in Figure 9-6a, and that is that all storage is at a single site. Should a disaster befall that site, then the entire system is down since all nodes are denied access to the database.

This condition can be corrected by adding a second NAS subsystem, as shown in Figure 9-6b, that is geographically remote from the first NAS subsystem. For active/active systems, the two NAS subsystems are preferably kept in synchronization via data replication. Now, if a storage site should fail, there is an alternate path to the data.

Redundant Unmirrored NAS

Again, with respect to Figure 9-6b, one may question whether four copies of the database (two mirrored pairs) are needed. One could instead configure each NAS subsystem with unmirrored disks (Figure 9-6c). This provides the network with dual database copies, which is generally sufficient to meet the requirements of extreme availability.

Mix and Match

In most applications, the need for data availability and access speeds may vary across the variety of databases used by the application or enterprise. Real-time applications may have need for extreme availability and fast access, but logs and archives may have significantly reduced availability and performance requirements. Therefore, consideration should be given to the use of a mix of these configurations to match the needs of each database application.

As mentioned earlier, RAID can be substituted as needed for single disks in order to achieve more reliability or for mirrored disks in order to reduce costs.

Recurring Costs

In addition to the initial costs for the system, there are many recurring costs that should be considered in the TCO equation.

Maintenance

The type of maintenance arrangement can have a direct impact on availability since availability is a function not only of the expected failure intervals for each component but also of its repair time. As repair time increases, there is a higher probability of a second failure while the first component is down. Repair times should be kept as short as is economically feasible.

There are several levels of maintenance to be considered. Fastest repair, justifiable for very large installations that are truly mission-critical, is achieved by having on-site maintenance personnel with an inventory of on-site equipment spares. The next most effective strategy is on-site spares with maintenance personnel available on short notice (say two hours).

Thereafter, the availability of spares and the response time of maintenance personnel are a matter of negotiation. However, the longer the repair cycle, the lower the availability. Extending the average repair time from two hours to one day reduces availability by one 9 (for example, from six 9s to five 9s).

Downtime

If an active/active system is being considered for implementation, the presumption is that any downtime is potentially quite painful and costly. The cost of downtime can result from lost business, lost customers, regulatory action, or a number of other reasons.

An extreme example of such a downtime cost was the failure of eBay's web site on June 10, 1999.[107] This 22-hour outage cost eBay somewhere between $3 million and $5 million in lost business. Perhaps even worse, eBay stock dropped by 25%, causing a loss of market capitalization in the billions of dollars. Its stock took nine months to recover.

eBay is not alone in such misfortunes. In August of 1996, an outage at AOL caused its stock to drop by 15%.[107] Catastrophic downtime will happen – it is just a matter of when. [108]

In a survey of 200 data center managers,[109] almost 80% of the respondents indicated that their cost of downtime was over $50,000 USD per hour Over 25% stated that it was over $500,000 USD per hour.

On the average, a system with four 9s availability is down almost an hour per year. If the cost of downtime is $100,000 USD per hour, this amounts to a downtime cost of $1,000,000 USD over a ten-year system life.

If the system has six 9s availability, the average yearly downtime is reduced by a factor of 100; and its cost is reduced to $1,000, or $10,000 over a ten-year period. If downtime carries with it a severe cost penalty, then the estimated cost of downtime must be included in TCO.

---

[107] Joseph Williams, *Avoiding the CNN Moment*, IEEE Computer Society Library; March/April,2001.
[108] As philosopher George Santanya once said, "Those who cannot remember the past are condemned to repeat it."
[109] USA Today, page 1B; July 3, 2006.

Of course, downtime can also carry some consequences not easily measured financially, such as customer dissatisfaction, bad publicity, loss of property or life, and legal liability.

Software

There are both software initial costs and recurring costs. The initial cost is, of course, the procurement or development of any specialized application software which might be required. Recurring costs include maintaining your software (discussed under personnel), software licenses, and software service charges.

Software licenses are often one of the largest of the recurring costs. Of particular importance is the vendor's policy for multinode systems. If licenses are insensitive to system size, then breaking a single system into a multinode active/active system can have large software license cost implications. A four-node system might have software license costs close to four times that of its monolithic system counterpart. Software licensing might be based on the number of nodes, the node size, and/or the applications running on the system. (However, licensing costs for an active/backup configuration are generally commensurate with single system licenses.)

Generally, software licensing models based on the amount of purchased total system capacity make the most sense when active/active architectures are being considered.

Beyond this, there are a number of time durations that might be selected. Perhaps licenses can be purchased outright, on a monthly or annual basis, or for a fixed term such as three years. The appropriate software licensing arrangements must be chosen. One may indeed find that different licensing agreements for different software modules are beneficial. These combinations can be analyzed by calculating their net present values or by doing an internal rate of return (IRR) analysis, as described earlier in this chapter.

When considering the total cost of ownership of a system, the number of nodes that might reflect the lowest cost may not always be obvious when hardware, software licenses, facilities, and people are considered. For instance, if 100% availability is to be provided in the event of a node failure, then 200% of capacity must be purchased if two nodes are to be used, 150% of capacity if three nodes are to be used, and 125% of capacity if four nodes are to be used. Thus, hardware costs may well decrease as the number of nodes increases. However, software licenses may increase; and facilities and personnel costs will definitely increase as the number of nodes increases. As a consequence, three nodes may be less expensive than two nodes; but four nodes may be more expensive than three nodes. Only a detailed TCO analysis can determine these relationships.

Another consideration is the capacity requirements for disaster recovery versus the capacity requirements for node failures. The system may be required to provide 100% of capacity in the event of a node failure; but in the unlikely event of a disaster, a reduced capacity may be deemed sufficient. Thus, a disaster recovery node might actually be a multinode campus environment. If a node on any campus fails, full capacity for the entire system can be maintained. However, should a disaster take out a campus, only reduced capacity would be available. By using such campus configurations, significant savings may be made in personnel and facilities costs.

Communications

Generally an active/active system will incur substantial communication costs. Only if all nodes are collocated and communicate over a local LAN are there no additional communication costs.

A variety of communication channel mechanisms can be chosen. They provide a tradeoff between cost, performance and reliability. At the bottom end is the Internet (typically implemented as a virtual private network, or VPN) – cheap but with relatively unreliable performance. At the top end are high-speed dedicated channels. These can be leased lines or ATM services.

In any event, redundancy must be considered. Should the primary data replication network go down, there must be a backup. Otherwise, connection loss can result in the so-called split-brain syndrome.[110] The Internet is a good candidate here because being packet switched, it is inherently reliable. Only the local connection to the Internet is of primary concern.

At the low cost end, communication backup can be dialed lines. At the high cost end, the backup may be a duplication of a private channel.

An important consideration is to provide the backup channel via a different carrier from the primary channel to avoid failure due to a carrier's massive outage. Furthermore, it should be determined that the two carriers do not use in their backbones some common carrier whose failure could take down both networks.

Network Management

We consider a network management facility separately from other software licensing not only because it can it be quite expensive but also because it can have a major impact on personnel and facilities costs.

As with any system, each node in an active/active system must be monitored for proper operation. This includes monitoring not only the processing and storage hardware but also the performance of the software modules, communication networks, system response times,

---

[110] See Chapter 3 of this volume, <u>Building Active/Active Systems</u>, Failure Mechanisms - Network Failure.

critical queues of work, and environmental factors such as power and air conditioning. In addition, there should be end-to-end monitoring to make sure that the application is functioning properly and that updates are, in fact, being properly replicated. Corrective action must be taken when needed.

The management of active/active systems is complicated by the fact that there are several nodes to monitor and that they may be geographically distributed. In addition, the nodes may be heterogeneous, comprising nodes of different versions of hardware or software; or they may even be provided by different vendors.

Excellent network management facilities that handle these tasks in a heterogeneous environment include HP's OpenView, BMC Patrol, Veritas Nerve Center, Tivoli Global Enterprise Manager, and CA Unicenter. These facilities can monitor and maintain a geographically dispersed network of nodes from a single location. They can reduce or perhaps eliminate the need for personnel at remote sites and as a consequence minimize facility costs as well.

However, such facilities can be quite expensive and can run into seven-figure costs in US dollars. Therefore, the purchase of a network management facility should be subject to its own ROI analysis. Its cost will depend on the breadth of services licensed. Its savings are the potential reduced personnel and facilities needed at remote sites as well as at the primary site and the decrease in downtime due to early warning of potential faults.

Personnel

Personnel costs can vary widely from system to system and from networking environment to networking environment. Areas to consider are:

- <u>System Administration</u> – Based on the administrative tools supplied with the system, different levels of staff support will be needed to administer a node and/or the entire system.

- <u>Database Administration</u> – The management of the database requires skills different from system administrators. Again, the facilities provided by the database often determine the level of database administrative support required. There are known cases of multiterabyte databases requiring only a few people for 24x7 support, whereas other databases of comparable size may require several times that number.

- <u>Software Support</u> – Any large application will require ongoing software maintenance to fix bugs, to tune performance, and to add enhancements.

- <u>Remote Nodes</u> – Depending upon the systems and the available network management facilities, each geographically distant node may have to be manned. This could range anywhere from full staffing at each node to a zero-staffing "lights out" operation for remote nodes.

For every critical position, there must be a backup person (in essence, a spare) to fill the role to cover vacations, sick leave, and so on.

A major potential investment in personnel is the need to staff remote nodes. This can be minimized (or eliminated) if a full network management facility is acquired and if fault-tolerant systems are used for the nodes. The additional costs of these approaches may be more than offset by the personnel savings that can be achieved in remote node staffing.

Facilities

Since the application must be distributed if disaster tolerance is to be achieved, remote facilities are a requirement. Facility distribution might be desirable from a locality of processing consideration or may be dictated by the need for disaster tolerance.

In any event, the cost of facilities, which include system and personnel space, desks and files, terminals, power, air conditioning, fire protection, raised flooring, and other environmental equipment, is dependent primarily on two factors – what does the system need and what do the people need.

The system facilities' requirements are dictated by the system. However, recent studies have shown that overconfiguring data processing center facilities can represent a major factor in the total cost of ownership. One particular study[111] showed an estimated cost of $120,000 per unused rack space (that is, the floor space required by one rack of equipment) over the lifetime of the data center. Over 30% of this cost can be saved by rightsizing the data center and letting it grow as needed.

Typically, the space deemed necessary in the long run must be initially acquired; and this is true of certain equipment associated with the data center such as that required for fire protection. However, many components, such as UPS capacity, power distribution, air conditioning, and raised flooring, can be added as needed in order to reduce the data center facility total cost of ownership.

The people requirements are strongly governed by the number of people that need to be supported at the node. Therefore, the steps that

---

[111] *Determining Total Cost of Ownership for Data Center and Network Room Infrastructure*, White Paper #6, APC (American Power Conversion); 2003.

can be taken to minimize personnel cost, especially at remote nodes as discussed in the previous section, can be an important factor in minimizing facilities' cost. As described above, these steps can include a good network management facility and fault-tolerant nodes to allow "lights out" operation at the remote sites.

Business Risk Insurance

Business risk insurance is designed to pool together the interests of a number of companies so that the premiums collected across a large number of businesses can be used to cover the losses of a single business. Business losses can be caused by natural disasters (floods, earthquakes, wind), fires, material or energy shortages, terrorist attacks, wars, worker strikes, foreign government nationalization, computer system failures, theft of real or intellectual property, cyber attacks, officer or director malfeasance, sabotage, product liability, and so on. However, many of these events are often excluded from business risk insurance.

An insurance premium can be thought of as comprising a basic coverage premium and a set of premiums for endorsed losses. The premium for an endorsement is a function of its risk, where risk can be thought of as

$$\text{risk} = \frac{\text{threat x vulnerability}}{\text{countermeasure}} \text{ x value}$$

The more expensive the system or the cost of its failure, the greater is the *value* and therefore the greater the risk to the insurance company. In terms described earlier, value includes not only replacement cost of equipment and facilities but also the cost of downtime. For this reason, the mean time to recover the system (its MTR) is of great importance.

Threat is the probability that the endorsed event will occur over the life of the policy. The time between endorsed events is the *mtbee* (mean time between endorsed events). In some cases, there is no control over mtbee – e.g., for floods or earthquakes. In other cases, mtbee can be controlled. For instance, operator errors can be minimized by proper training, documentation, and support utilities.

The system *vulnerability* is the probability that a threat could cause a loss. Vulnerability could be the location of the system on a fault line, in a hurricane or tornado zone, or in a locale in which the current political climate is unstable.

The *countermeasure* is what has been done to minimize the risk. This is where active/active systems come in. An active/active system is an effective countermeasure against many endorsed threats, and it can reduce the value of an event by providing very fast recovery time. Therefore, it can significantly lower insurance costs.

Rule 56: *To minimize insurance costs, reduce threat by increasing mtbee (the mean time between endorsed events), reduce vulnerability by using appropriate countermeasures, and minimize the value of an event by minimizing MTR. An active/active system is an effective countermeasure that reduces both vulnerability and event value.*

Various governmental regulations may address the structure of or requirements for business risk insurance in certain industries. In other industries, business risk insurance may not even be available or may be prohibitively expensive.

Certain catastrophic events such as earthquakes, floods, and war may not be covered at all. Severe losses in an industry could exhaust the insurance pool, in which case further losses may not be covered.

It is quite possible that redundancy could reduce premiums because of the reduced risk of business loss. An active/active geographically distributed system may enjoy significant reductions in business risk insurance premiums in certain cases.

In any event, a well-thought-out and well-practiced disaster recovery plan will often go a long way toward minimizing the potential for business loss and therefore the premiums of business risk insurance.

A TCO Toolkit

The proper calculation of TCO for a system can be a daunting task. There are, however, services available to help. One such service is VirtualADVISOR from The Standish Group.[112]

VirtualADVISOR uses a set of Standish tools which accept a wide range of input parameters from the client concerning the costs of the current system and the proposed system. The VirtualADVISOR tools are then used to calculate various TCO and ROI parameters.

The process begins with the VirtualADVISOR client working with a Standish STAR (STandish AdvisoR). This experienced field operative guides the client through the process of entering the required input data into a VirtualADVISOR worksheet. The data covers the factors that we have already discussed, including the following costs for both the current system and the proposed system:

- hardware
- software licenses
- manpower
- hardware and software maintenance
- facilities

---

[112] See www.standishgroup.com

- software infrastructure
- database and system administration
- application maintenance
- migration
- other (trade-ins, credit, and so forth)

The input tool allows any miscellaneous category to be defined and cost factors entered for that category. Various revenue categories may also be defined and entered.

Information about the applications is entered, including such items as the type of application (transaction, batch, query), transaction rates, both the current and proposed systems, and the current and proposed databases.

This input data is then passed to a Standish in-house specialist. Using a sophisticated project database called CHAOS and other tools, the expected range of each cost element is estimated; and optimum solutions are created. Standish uses a process which they describe as case-based reasoning technology driven by the CHAOS database to measure and profile the input data. The CHAOS database contains over 100 pertinent attributes from over 50,000 cases submitted by over 5,000 project managers through surveys, user interviews, vendor interviews, and focus workshops.

At this point, ROI and TCO tools are used to calculate a variety of cost and return-on-investment parameters. The results include for a specified time period:

- cost of implementing and operating the system
- capitalized savings
- operational savings
- cumulative gain (savings – cost)
- return on investment
- payback period

371

The in-house specialist then prepares a report explaining these results and defining the best strategy to maximize ROI. The report is passed to the client's STAR, who works with the client to review the outcomes and to make recommendations for paths to pursue.

VirtualADVISOR requires neither training nor any installation of software. The client's interface is strictly through his STAR.

Though not related to TCO and ROI, another major part of the VirtualADVISOR service is the assessment of risk associated with implementing a project. This is supported by the extensive case study attributes found in the CHAOS database.

Putting It All Together

As we have seen, TCO is a complex combination of many factors. As with any system, TCO starts with the initial costs. However, active/active systems provide many configuration options with respect to the number of sites, the number of nodes, the number of processors at each node, the processors, the networks, and the disk subsystems. The proper configuration must be chosen to provide the optimal compromise between availability, cost, and performance. The acceptable level of downtime relative to your firm's Recovery Point Objective (RPO) and Recovery Time Objective (RTO)[113] is also an important factor to be considered in the cost of the system.

Next, a network management facility must be chosen. The capabilities of this facility coupled with the reliability of each node (which is certainly enhanced by the use of fault-tolerant nodes) will determine the staffing required at each node. To the extent that a "lights out" operation can be achieved at a node, personnel and facilities costs can be minimized.

---

[113] See Volume 1, Chapter 6, RPO and RTO.

An important observation from these discussions is that the use of fault-tolerant nodes can have several positive impacts on cost:

- Fewer nodes may have to be purchased.
- Software licensing costs may possibly be lower.
- Fewer nodes mean fewer facilities.
- Fewer facilities mean less personnel.
- Personnel requirements may be further reduced at remote nodes since immediate maintenance is not such a concern.

This leads us to the following:

Rule 57: (Corollary to Rule 39) - *The use of fault-tolerant nodes in an active/active system can reduce TCO by reducing the number of nodes required, the facilities' costs, the licensing costs, and the personnel costs.*

It is often the perception that active/active systems have a higher TCO until cost of downtime is considered. A careful TCO study will often show otherwise. It behooves management to carefully consider all options to come up with the most economical system measured over its life span. Only in this way can ROI truly be optimized.

What's Next

This chapter concludes Volume 2 of the *Breaking the Availability Barrier* series.

We speak often of the triumvirate of availability, cost, and performance. We have explored availability in Chapter 1 and cost in this chapter. We move now to Volume 3 and turn our attention to performance.

The performance of an active/active system is largely determined by two factors – the asynchronous replication latency or the synchronous application latency, as appropriate,[114] of the data replication engine; and the transit time of the communication links (largely out of our control as they are governed by the speed of light).

In our next chapter, Chapter 10 of Volume 3, we evaluate active/active system performance for the two primary approaches that we have discussed – data replication and network transactions. This analysis is highly mathematical, but the mathematics are relegated to Appendix 2 of Volume 3. Chapter 10 primarily summarizes the lessons learned from this analysis – no math required.

---

[114]See Volume 1, Chapter 3, <u>Asynchronous Replication</u> and Chapter 4, <u>Synchronous Replication</u>.

Appendices

Throughout Volumes 1 and 2 and upcoming in Volume 3, we have formulated rules as a way to highlight critical points. These are summarized in Appendix 1. The Volume and Chapter in which each rule was formulated is noted for ready reference to aid the review of the context for that rule.

Appendix 1 – Rules of Availability

"Controversial proposals, once accepted, soon become hallowed."
- Dean Acheson

In Volumes 1, 2, and 3 of this series, we have described a number of rules that have to do with highly available systems. These form a useful review of the important issues and results that we have described. These rules are summarized below, organized by volume, and indexed by chapter for ready reference.

Volume 1 Rules

Chapter 1 – The 9s Game

Rule 1: *If all subsystems must be operational, then the availability of the system is the product of the availabilities of the subsystems.*

Rule 2: *Providing a backup doubles the 9s.*

Rule 3: *System reliability is inversely proportional to the number of failure modes.*

Rule 4: *Organize processors into pairs, and allocate each process pair only to a processor pair.*

Rule 5: *If a system can withstand the failure of s subsystems, then the probability of failure of the system is the product of the probability of failures of (s+1) systems.*

Rule 6: *System availability increases dramatically with increased sparing. Each additional level of sparing adds a subsystem's worth of 9s to the overall system availability.*

Rule 7: *For a single spare system, the system MTR is one-half the subsystem mtr.*

Rule 8: *For the case of a single spare, cutting subsystem mtr by a factor of k will reduce system MTR by a factor of k and increase the system MTBF by a factor of k, thus increasing system reliability by a factor of k^2.*

Chapter 2 – System Splitting

Rule 9: *If a system is split into k parts, the resulting system network will be more than k times as reliable as the original system and still will deliver (k-1)/k of the system capacity in the event of an outage.*

Rule 10: *If a system is split into k parts, the chance of losing more than 1/k of its capacity is many, many times less than the chance that the single system will lose all of its capacity.*

Chapter 3 – Asynchronous Replication

Rule 11: *Minimize data replication latency to minimize data loss following a node failure.*

Rule 12: *Database changes generally must be applied to the target database in natural flow order to prevent database corruption.*

Rule 13: *Follow natural flow order when replicating so as not to create artificial activity peaks at the target database.*

Rule 14: *Block the ping-ponging of data changes in a bidirectional replication environment to prevent database corruption.*

Rule 15: *Minimize replication latency to minimize data collisions.*

Rule 16: (Gray's Law) - *Waits under synchronous replication become data collisions under asynchronous replication.*

Chapter 4 – Synchronous Replication

Rule 17 (as enhanced in Chapter 4 of Volume 2): *For synchronous replication, coordinated commits using data replication become more efficient relative to network transactions and replicated lock management under a transaction manager as transactions become larger, as communication channel propagation time increases, or as the transaction load increases.*

Chapter 5 – The Facts of Life

Rule 18 : *Redundant hardware systems have an availability of five to six nines. Software and people reduce this to four nines or less.*

Rule 19: (Bartlett's Law) - *When things go wrong, people get stupider.*

Rule 20: *Conduct periodic simulated failures to keep the operations staff trained and to ensure that recovery procedures are current.*

Rule 21: *System outages are predominantly caused by human and software errors.*

Rule 22: (Corollary to Rule 20) - *A system outage usually does not require a repair of any kind. Rather, it entails a recovery of the system.*

Rule 23: (Niehaus' Law) - *Change causes outages.*

Rule 24: *Following the failure of one subsystem, failover faults cause the system to behave as if it comprises n-1 remaining subsystems with decreased availability.*

Rule 25: *The possibility of failover faults erodes the availability advantages of system splitting (see Rule 9).*

Rule 26: (The Golden Rule) - *Design your systems for fast recovery to maximize availability, to reduce the effect of failover faults, and to take full advantage of system splitting.*

Rule 27: *Rapid recovery of a system outage is not simply a matter of command line entries. It is an entire business process.*

Chapter 6 – RPO and RTO

Rule 28: *RPO and RTO are both a function of the data replication technology used to maintain databases in synchronism.*

Chapter 7 – The Ultimate Architecture

Rule 29: *You can have high availability, fast performance, or low cost. Pick any two.*

Rule 30: *A system that is down has zero performance, and its cost may be incalculable.*

Chapter 9 – Data Conflict Rates

Rule 31: *Minimize lock latency to minimize synchronous replication deadlocks.*

Rule 32: *Lock latency deadlocks under synchronous replication become collisions under asynchronous replication.*

Rule 33: *Designating a master node for lock coordination can eliminate lock latency deadlocks when using synchronous replication.*

Chapter 10 – Referential Integrity

Rule 34: *Database changes must be applied to the target database in natural flow order to maintain referential integrity.*(See Rule 12.).

Rule 35: *A serializing facility that will restore natural flow is required following all data replication threads and before the target* database in order to guarantee that the database will remain consistent and uncorrupted.

Volume 2 Rules

Forward

Rule 36: *To achieve extreme reliabilities, let it fail; but fix it fast.*

Chapter 1 – Achieving Century Uptimes

Rule 37: *All of the purchased capacity in an active/active system is usable. There is no need for an idle standby system.*

Rule 38: *Providing that the nodes in an active/active system are geographically distributed, disaster tolerance comes for free.*

Rule 39: (NonStop Maxim) - *Century uptimes in an active/active system will generally require fewer nodes if fault-tolerant nodes are used rather than high availability nodes.*

Rule 40: (Darwin's extension to Murphy's Law) - *Eventually, a disaster will befall every enterprise; and only those that are prepared will survive.*

Chapter 2 – Reliability of Distributed Computing Systems

Rule 41: *Active/active systems can provide the availability of a primary/standby pair with less equipment and less cost.*

Rule 42: *In a system with s spares, reducing subsystem mtr by a factor of k will reduce system MTR by a factor of k and will increase system MTBF by a factor of k^s, thus increasing system reliability by a factor of k^{s+1}.*

Rule 43: *If s+1 subsystems fail and are being repaired simultaneously, and if the return to service of any one of these* subsystems will return the system to service, the system MTR is *mtr/(s+1)*.

Rule 44: (Corollary to Rule 42) - *In a system with s spares, going to parallel repair will decrease MTR by a factor of (s+1), will increase MTBF by a factor of s!, and will decrease the probability of failure by a factor of (s+1)!.*

Rule 45: *If you break a monolithic system into k smaller nodes with no spares, the system will be more reliable than the original monolithic system provided that each node is more than k times as reliable as the monolithic system.*

Rule 46: *Don't underestimate your failover time. It may well be the most important factor in perceived availability.*

Rule 47: *A small probability of a failover fault may cause a disproportionate decrease in system availability. Moving from active/backup to active/hot-standby or active/active with frequent testing can significantly reduce the probability of failover faults.*

Rule 48: *Pick your node locations in an active/active system carefully to minimize the chance that environmental hazards will outweigh the availability of the nodes.*

Chapter 3 – Building Active/Active Systems

Rule 49: *Replication engines that violate referential integrity on the target database are rarely, if ever, suitable for active/active implementations.*

Chapter 4 - Active/Active and Related Topologies

Rule 50: (Corollary to Rule 46) - *Hardware replication is not suitable for active/active applications because it does not provide referential integrity.*

Chapter 5 – Redundant Reliable Networks

Rule 51: (Corollary to Rule 2) - *Providing a redundant network doubles the 9s of the network.*

Rule 52: (Corollary to Rule 9) - *Two smaller networks are more reliable than one large network.*

Chapter 8 – Eliminating Planned Outages with Zero Downtime Migrations (ZDM)

Rule 53: *ZDM (zero downtime migrations) eliminates planned application downtime, therefore improving application availability.*

Chapter 9 – Total Cost of Ownership (TCO)

Rule 54: (Standish Law) - *In order to calculate a meaningful predicted Total Cost of Ownership, one must first properly size the system.*

Rule 55: *Remember the power of negotiation. For instance, at the end of a vendor's fiscal quarter, sales goals may drive significant discounts.*

Rule 56: *To minimize insurance costs, reduce threat by increasing mtbee (the mean time between endorsed events), reduce vulnerability by using appropriate countermeasures, and minimize the value of an event by minimizing MTR. An active/active system is an effective countermeasure that reduces both vulnerability and event value.*

Rule 57: (Corollary to Rule 39) - *The use of fault-tolerant nodes in an active/active system can reduce TCO by reducing the number of nodes required, the facilities' costs, the licensing costs, and the personnel costs.*

Volume 3 Rules

Chapter 10 – Performance of Active/Active Systems

Rule 58: (Latency Rule) - *Replication engine latency is largely governed by disk queuing and communication delays, not by processing times.*

Rule 59: (Throughput Rule) - *If the source and target systems are geographically separated, the throughput of the data replication engine is likely to be dominated by the communication channel. If the systems are collocated, the Appliers are likely to be the dominant factor.*

Rule 60: (Fool's Rule) - *If transaction response time is much longer than the application latency due to other factors, there is no apparent advantage to either coordinated commits or network transactions for synchronous replication (NOT!).*

Chapter 14 – Benefits of Multiple Nodes in Practice

Rule 61: *Active/active systems made up of fault-tolerant nodes will benefit from lights-out operations. Active/active systems using high-availability (but not fault-tolerant) nodes may have, in fact, poorer availability in a lights-out operation than the individual nodes.*

Appendix 2 – Replication Engine Performance Model

Rule 62: (Paul's Law) - *If synchronous replication is used, the Applier does not have to enforce the order of commits so long as the source referential integrity is at least as strong as that at the target.*

Rule 63: *The difference in application latency between coordinated commits and dual writes is most apparent when fast response times are required.*

Appendix 4 – A Consultant's Critique

Rule 64: (Werner's Law) - *When trying to distribute an application over multiple nodes, identify each global resource and local context the application uses; and carefully consider the consequences of this in a multinode approach. Identify each application decision which is made on system or database state information (as opposed to database contents information), and make sure this is compatible with your multinode approach.*

References and Suggested Reading

The following references include those mentioned in Volume 1 and Volume 2 of this series. They also include some additional references that are quite useful for further background material.

Advanced Computer and Network Corporation, *RAID 6*, www.acnc.com.

Availability Digest, www.availabilitydigest.com.

Barker, R., Massiglia, P., *Storage Area Network Essentials*, Wiley Computer Publishing; 2002.

Bartlett, W., *Indestructible Scalable Computing*, ITUG Summit presentation; September, 2001.

Brodie, M. L., Stonebraker, M., *Migrating Legacy Systems: Gateways, Interfaces, & the Incremental Approach,* Morgan Kaufmann Publishers; 1995.

Buckle, R., Highleyman, W. H., *The New NonStop Advanced Architecture: A Massive Jump in Processor Reliability*, The Connection, Volume 24, Number 6; September/October, 2003.

Chien, A., article in *The Grid2: Blueprint for a New Computing Architecture,* a collection of articles edited by Ian Foster and Carl Kesselman, Morgan Kaufmann; 2004.

Compaq Computers, *Disaster Tolerance: The Technology of Business Continuity*, www.techguide.com.

Devraj, V. S., *Oracle 24x7 Tips & Techniques*, <u>Oracle Press</u>; 2000.

Einhorn, S. J., *Reliability Prediction for Repairable Redundant Systems*, <u>Proceedings of the IEEE</u>; February, 1963.

Faithful, M., *The a-BCP of DR*, <u>The Connection</u>, Issue 23, Number 6; November/December, 2002.

FDBS Characteristics: www-128.ibm.com/developerworks/db2/library/techarticle/0203haas

Fox, A., Patterson, D., *Self-Repairing Computers*, <u>Scientific American</u>; June, 2003.

Gray, J., *Why Do Computers Stop and What Can We Do About It?* <u>5th Symposium on Reliability in Distributed Software and Database Systems</u>; 1986.

Gray, J., Reuter, A., *Transaction Processing: Concepts and Techniques*, Morgan Kaufmann; 1993.

Gray, J., Helland, P., O'Neil, P., Shasha, D., *The Dangers of Replication and a Solution,* <u>ACM SIGMOD Record</u> (Proceedings of the 1996 ACM SIGMOD International Conference on Management of Data), Volume 25, Issue 2; June, 1996.

Grid white papers: <u>www.ggf.org/ogsi-wg</u>.

Hewlett-Packard, *HP NonStop servers ranked tops in availability by Standish, beating IBM Sysplex and RS/6000 Non-Clusters*, www.himalaya.compaq.com/object/STANTOPSNW; September, 2002.

Highleyman, W. H., *Performance Analysis of Transaction Processing Systems,* Prentice-Hall; 1988.

Highleyman, W.H., *Distributing OLTP Data Via Replication*, <u>The Connection</u>, Volume 16, No. 2; April/May, 1995.

Highleyman, W. H., *Reliability Analysis – It's More Important Than Ever*, The Connection, Volume 22, Number 3; May/June, 2001.

Highleyman, W. H., *The Impact of Mean Time to Repair on System Availability*, ITI, Inc., White Paper, August 19, 2002.

Highleyman, W. H., Holenstein, P. J., Holenstein, B. D., Six-Part Availability Series, The Connection:
  Availability (Part 1) – The Nines Game; Nov./Dec., 2002.
  Availability (Part 2) – System Splitting; Jan./Feb., 2003.
  Availability (Part 3) – Asynchronous Replication; March/April, 2003.
  Availability (Part 4) – Synchronous Replication; May/June, 2003.
  Availability (Part 5) – The Ultimate Architecture; Sept./Oct., 2003.
  Availability (Part 6) – RPO and RTO; March/April, 2004.

Highleyman, W. H., Holenstein, P. J., Holenstein, B. D., *Achieving Century Uptimes* Series, The Connection:
  Part 1: Survivable Systems for Enterprise Computing; Nov./Dec., 2006.
  Part 2: What Will Active/Active Cost Me?; Jan./Feb., 2007.
  Part 3: Avoiding Data Collisions; March/April, 2007.
  Part 4: Detecting and Resolving Data Collisions; May/June, 2007.

Highleyman, W. H., Holenstein, P. J., Holenstein, B. D., *Method of Increasing System Availability by Splitting a System*, United States Patent 7,113,938; September 26, 2006.

Highleyman, W. H., Holenstein, P. J., Holenstein, B. D., *Method of Increasing System Availability by Assigning Process Pairs to Processor Pairs*, United States Patent Application 10/367,675; February 13, 2003.

389

Highleyman, W. H., Holenstein, P. J., Holenstein, B. D., *Breaking the Availability Barrier: Survivable Systems for Enterprise Computing*, AuthorHouse; 2004.

Highleyman, W. H., Holenstein, B. D., Holenstein, P. J., *The Availability Corner,* <u>The Connection</u>:
>   *Let's Measure System Reliability in Centuries;* Nov./Dec., 2004.
>   *What Reliability Do We Really Need?*; Jan./Feb., 2005.
>   *The Language of Availability;* May/June, 2005.
>   *The Great Tape Backup Paradigm Shift;* July/Aug., 2005.
>   *Fault Tolerance vs. High Availability;* Sep./Oct., 2005.
>   *TCO for Active/Active Systems;* Nov./Dec., 2005.
>   *The Net Present Value of Active/Active Systems;* Jan./Feb., 2006.
>   *Grid Computing;* March/April, 2006.
>   *Is IBM's Parallel Sysplex a NonStop Competitor?*; July/Aug., 2006.
>   *Testing Your System Recovery Plan*; Sept./Oct., 2006.

Holenstein, B. D., Waterstraat, M., *Integrating Platforms in the E-Commerce Enterprise*, <u>The Connection</u>, Volume 21, Number 4; July/August, 2000.

Holenstein, P. J., Holenstein, B. D., *High-Availability Web Site/NSK Cooperative Processing Using Database Replication in a ZLE Architecture*, <u>The Connection</u>, Volume 22, Number 5; September/October, 2001.

Holenstein, B. D., Holenstein, P. J., Highleyman, W. H., *Asynchronous Coordinated Commit Replication and Dual Write with Replication Transmission and Blocking of Target Database Updates Only*, <u>United States Patent Application US2004/0133591A1</u>; September 17, 2003.

Holenstein, B. D., Holenstein, P. J., Strickler, G. E., *Collision Avoidance in Bidirectional Database Replication*, <u>United States Patent 6,662,196</u>; December 9, 2003.

Holenstein, B. D., *2003 HP User Groups BC-DR Survey Responses*, <u>The Connection</u>; Jan./Feb., 2004.

Holenstein, P. J., Holenstein, B. D., Highleyman, W. H., *Method for Ensuring Referential Integrity in Multithreaded Replication Engines,* United States Patent Application 10/881,699; June 30, 2004.

Holenstein, P. J., Holenstein, B. D., Strickler, G. E., *Synchronization of Plural Databases in a Database Replication System,* United States Patent 6,745,209; June 1, 2004.

Holenstein, P. J., Holenstein, B. D., Strickler, G. E., *Synchronization of Plural Databases in a Database Replication System,* United States Patent 7,003,531; February 21, 2006.

Holenstein, B. D., Holenstein, P. J., Highleyman, W. H., *Asynchronous Coordinated Commit Replication and Dual Write with Replication Transmission and Blocking of Target Database Updates Only,* United States Patent 7,177,866; February 13, 2007.

HP User Advocacy Survey and Instapoll results posted on www.hpuseradvocacy.org. on October 2, 2003, and on December 18, 2003.

IBM Corporation, *Parallel Sysplex Availability Checklist*; May, 2003.

ITUG Disaster Recovery SIG,
http://www.itug.org/sigs/disaster.cfm?
CFID=34497310&CFTOKEN=71445435

Keeping Systems Running, Connexion 2; September 1992.

Kleinrock, L., *Queuing Systems,* Vol. 2, Computer Applications, John Wiley and Sons; 1976.

Knapp, H. W., *The Natural Flow of Transactions,* ITI, Inc,. White Paper; 1996.

Lan Downtime Clear and Present Danger, Data Communications; March 21, 1990.

LaPedis, R., *Developing Contingency Plans for the Recovery of Critical Data and Applications,* Compaq White Paper; 1999.

LaPedis, R., *RTO and RPO Not Tightly Coupled,* Disaster Recovery Journal; Summer, 2002.

LaPedis, R., *Will Enterprise Storage Replace NonStop RDF?* The Connection, Volume 23, Issue 6; November/December, 2002.

Liebowitz, B. H., Carson, J. H., Chapter 8 – Reliability Calculations, *Multiple Processing Systems for Real-Time Applications*, Prentice-Hall; 1985.

McNamara, J., *Technical Aspects of Data Communication*, Digital Press; 1988.

Marcus, E., Stern, H., *Blueprints for High Availability: Designing Resilient Distributed Systems*, Wiley Computer Publishing; 2000.

Merian, L., *Bulletproof Storage*, Computerworld; April 13, 2005.

Microsoft's SQL Server Distributed Partitioned Views msdn.microsoft.com/library/en-us

Open Grid Service Infrastructure Primer, www.ggf.org/ogsi-wg; August 25, 2004.

Open Grid Services Architecture Use Cases, forge.gridforum.org/projects/ogsa-wg; October 26, 2004.

Oracle White Paper, *Database Architecture: Federated vs. Clustered*; February, 2002.

Pfister, G. F., *In Search of Clusters: The Ongoing Battle In Lowly Parallel Computing*, Prentice-Hall PTR; 1998.

Shannon, T. C., *HP Brings Enterprise-Class Storage Capabilities to the Midrange,* Shannon Knows HPC, Volume 10, Issue 18; April 15, 2003.

Shannon, T. C., *HP Grapples With the Grid, NonStop MIA From HP Grid Strategy,* Shannon Knows HPC, Volume 11, Issue 14; March 10, 2004.

Stevens, W., *TCP/IP Illustrated, Volume1 – The Protocols,* Addison-Wesley; 1994. (See especially Chapters 2, 9, 10, 17.)

Strickler, G. E., Knapp, H. W., Holenstein, B. D., Holenstein, P. J., *Bidirectional Database Replication Scheme for Controlling Ping-Ponging,* United States Patent 6,122,630; Sept. 19, 2000.

Taneja, A., *Perpetual Motion Enterprise,* Intelligent Enterprise; January 14, 2002.

The Standish Group, *The New High Availability – A NonStop Continuous Processing Architecture (CPA),* Standish Group Research Note; 2002.

The Standish Group, VirtualBEACON, Issue 244; September, 2002.

Thornburgh, R. H., Schoenborn, B. J., *Storage Area Networks,* Prentice Hall PTR; 2001.

Torres-Pomales, W., *Software Fault Tolerance: A Tutorial,* NASA White Paper; October, 2000.

Weiss, T. R., *Giving Bugs the Boot,* ComputerWorld; November 8, 2004.

Weygant, P. S., *Clusters for High Availability,* Prentice Hall PTR; 2001.

Wikipedia Encyclopedia, en.wikipedia.org, for instance:

availability
data replication
federated database
grid computing
internal rate of return
net present value
Parallel Sysplex
present value
recovery point objective
recovery time objective
referential integrity
synchronous replication
unavailability
virtual tape

Williams, J., *Avoiding the CNN Moment*, IEEE Computer Society Library; March/April, 2001.

Wong, B. L., *Configuration and Capacity Planning for Solaris Servers*, Prentice-Hall; 1997.

Wood, A., A*vailability Modeling*, Circuits and Devices; May, 1994.

Wood, A, *Predicting Client/Server Availability*, Computer, Volume 28, Number 4; April, 1995.

Worthington, D., *Interview: The Future in Grid Computing*, BetaNews; February 21, 2005.

Vendor Web Sites:
- www.gravic.com.
- www.hp.com/go/nonstop
- www.netweave.com.
- www.standishgroup.com
- www.taplabs.com

Index

About the Authors

Paul J. Holenstein, Dr. Bill Highleyman, and Dr. Bruce Holenstein have a combined experience of over 90 years in the implementation of fault-tolerant, highly available computing systems. This experience ranges from the early days of custom redundant systems to today's fault-tolerant offerings from HP (NonStop) and Stratus.

Paul J. Holenstein and Dr. Bruce Holenstein have architected and implemented the various data replication techniques required for the active/active availability enhancements described in this book. Their company, Gravic, provides the Shadowbase line of data replication products to the fault-tolerant community. They hold several patents in the field of data replication.

Dr. Bill Highleyman has done extensive work on the effect of failure-mode reduction on system availability. He has built fault-tolerant systems for train control, racetrack wagering, securities trading, message communications, and other applications. He has published extensively on availability, performance, middleware, and testing. He is author of *Performance Analysis of Transaction Processing Systems*, published by Prentice-Hall and is Managing Editor of the *Availability Digest*, a monthly periodical focusing on the quest for 100% uptime (www.availabilitydigest.com).

All three authors also coauthored the book *Breaking the Availability Barrier: Survivable Systems for Enterprise Computing*, ISBN 1-4107-9233-1, which lays a great deal of groundwork for the material presented in this book.

CPSIA information can be obtained at www.ICGtesting.com
Printed in the USA
BVOW03s0114210415

396949BV00001B/100/P